Lecture Notes of the Institute for Computer Sciences, Social Informatics and Telecommunications Engineering

554

The LNICST series publishes ICST's conferences, symposia and workshops.
  LNICST reports state-of-the-art results in areas related to the scope of the Institute.
  The type of material published includes

- Proceedings (published in time for the respective event)
- Other edited monographs (such as project reports or invited volumes)

  LNICST topics span the following areas:

- General Computer Science
- E-Economy
- E-Medicine
- Knowledge Management
- Multimedia
- Operations, Management and Policy
- Social Informatics
- Systems

Jingchao Li · Bin Zhang · Yulong Ying
Editors

6GN for Future Wireless Networks

6th EAI International Conference, 6GN 2023
Shanghai, China, October 7–8, 2023
Proceedings, Part II

 Springer

Editors
Jingchao Li (ID)
Shanghai Dianji University
Shanghai, China

Bin Zhang
Kanagawa University
Yokohama, Japan

Yulong Ying (ID)
Shanghai University of Electric Power
Yangpu, China

ISSN 1867-8211 ISSN 1867-822X (electronic)
Lecture Notes of the Institute for Computer Sciences, Social Informatics
and Telecommunications Engineering
ISBN 978-3-031-53403-4 ISBN 978-3-031-53404-1 (eBook)
https://doi.org/10.1007/978-3-031-53404-1

This Springer imprint is published by the registered company Springer Nature Switzerland AG
The registered company address is: Gewerbestrasse 11, 6330 Cham, Switzerland

Paper in this product is recyclable.

Preface

We are delighted to introduce the proceedings of the sixth edition of the European Alliance for Innovation (EAI) International Conference on 6G for Future Wireless Networks (6GN 2023). This conference brought together researchers, developers and practitioners around the world who are leveraging and developing smart grid technology for a smarter and more resilient grid. The theme of EAI 6GN 2023 was "Smart Grid Inspired Future Technologies: A smarter and more resilient grid".

The technical program of EAI 6GN 2023 consisted of 60 full papers, including 3 invited papers in oral presentation sessions at the main conference tracks. The conference tracks were: Track 1 - 6G Communications for UAVs; Track 2 - Signal Gene Characteristics; Track 3 - Signal Gene Characteristics. Aside from the high-quality technical paper presentations, the technical program also featured three invited keynote speeches. The first invited talk was presented by Cesar Briso from Technical University of Madrid, Spain. The title of the talk was A Connected Sky: 6G Communications for UAVs. The second invited talk was presented by Peng Chen from Southeast University, China. The title of the talk was Efficient DOA Estimation Method for Reconfigurable Intelligent Surfaces Aided UAV Swarm. The third invited talk was presented by Jingchao Li from Shanghai Dianji University, China. The title of the talk was Physical layer authentication method for the Internet of Things based on radio frequency signal gene.

Coordination with the steering chairs, Imrich Chlamtac, Victor C.M. Leung and Kun Yang, was essential for the success of the conference. We sincerely appreciate their constant support and guidance. It was also a great pleasure to work with such an excellent organizing committee team for their hard work in organizing and supporting the conference. In particular, the Technical Program Committee completed the peer-review process of technical papers and made a high-quality technical program. We are also grateful to all the authors who submitted their papers to the EAI 6GN 2023 conference and workshops.

We strongly believe that EAI 6GN 2023 provided a good forum for all researchers, developers and practitioners to discuss all science and technology aspects that are relevant to wireless networks. We also expect that 6GN conferences will be as successful and stimulating as indicated by the contributions presented in this volume.

Yulong Ying

Organization

Organizing Committee

General Chairs

Junjie Yang	Shanghai Dianji University, China
Yulong Ying	Shanghai University of Electric Power

General Co-chairs

Cheng Cai	Shanghai Dianji University, China
Yulong Ying	Shanghai University of Electric Power, China

TPC Chair and Co-chairs

Yudong Zhang	University of Leicester, UK
Peng Chen	Southeast University, China
Jingchao Li	Shanghai Dianji University, China
Wanying Shi	Portland State University, USA
Pengpeng Zhang	Shanghai Dianji University, China
Na Wu	Nanjing University of Posts and Telecommunications, China

Sponsorship and Exhibit Chair

Siyuan Hao	Qingdao University of Technology, China

Local Chair

Ming Li	Shanghai Dianji University, China

Workshops Chairs

Haijun Wang	Shanghai Dianji University, China
Zhimin Chen	Shanghai Dianji University, China

Publicity and Social Media Chair

Tingting Sui   Shanghai Dianji University, China

Publications Chairs

Bin Zhang   Kanagawa University, Japan
Ao Li   Harbin University of Science and Technology,
                                   China

Web Chair

Xiaoyong Song   Shanghai Dianji University, China

Posters and PhD Track Chair

Yang Xu   Guizhou University, China

Panels Chair

Shuihua Wang   University of Leicester, UK

Demos Chair

Yue Zeng   Jinling Institute of Technology, China

Tutorials Chair

Pengyi Jia   Western University, Canada

Technical Program Committee

Jingchao Li   Shanghai Dianji University, China
Bin Zhang   Kanagawa University, Japan
Peng Chen   Southeast University, China

Contents – Part II

**Communications Systems and Networking & Control and Automation
Systems**

Computer Systems and Applications

Contents – Part I

Artificial Intelligent Techniques for 6G Networks

Power and Energy Systems

Image, Video, and Signal Processing

Detection of Green Walnuts on Trees Using the Improved YOLOv7 Model

Jinrong He[✉], Yuanhao Liu, Longlong Zhai, and Hanchi Liu

College of Mathematics and Computer Science, Yan'an University, Yan'an 716000, China
hejinrong@yau.edu.cn

Abstract. One application of artificial intelligence in agriculture is the use of machines to detect fruits and evaluate yield. Due to the small size and color of green walnuts similar to leaves, it is important to develop a method that detects walnuts quickly and accurately. Motivated by this issue, we propose a solution using the improved YOLOv7 model and use the improved model for detection and identification. We constructed a dataset with data augmentation to help with this study, containing a total of 10,550 images, including green walnuts from different angles. We used Precision, Recall, F-Measure, and mean Average Precision as the accuracy indexes of the model. Add the Transformer model, the ResNet network, and the SimAm attention mechanism to the network structure of the YOLOv7 model to improve the detection capability of the model. Compared to the YOLOv7 model without improvements, P increased by 1.5%, R increased by 1.3%, F1 increased by 1%, and mAP increased by 1.5%. Compared with other target detection models, the accuracy indexes show better results. This method can maintain high precision in walnut identification and detection and can provide technical support to the machine to recognize walnuts in a complex environment quickly and for a long time. The dataset is publicly available on Github: https://github.com/lunchc/Walnut.

Keywords: green walnut · YOLOv7 · object detect · deep learning · computer vision

1 Introduction

With the development of technology, artificial intelligence is widely used in agriculture. Intelligence and accurate crop management techniques have drawn the wide of attention science, and they can reduce labor consumption and increase the economic efficiency of an orchard.

In China, the walnut is one of the most widely distributed economic tree species. The plant area and walnut output are ranked first in the world. And the walnut have played an important role in increasing farmers income, alleviating industrial poverty, and ecological construction [1]. The reason is that the color of the unpeeled walnut is similar to that of the leaves and small in size. Detecting green walnuts is more difficult

J. Li et al. (Eds.): 6GN 2023, LNICST 554, pp. 3–17, 2024.
https://doi.org/10.1007/978-3-031-53404-1_1

than detecting other fruits. Therefore, detecting the green walnut is a great challenge for computer vision.

In recent years, the machine vision research in agricultural field detection, with the development of artificial intelligence, has been rapidly updated. Using traditional machine vision methods to detect walnuts requires designers to extract key characteristics such as size, color [2, 3], shape [4, 5], and texture [6, 7]. Then, the key features are put into the machine learning model, such as the support vector machine [8, 9], the decision tree [10, 11], the artificial neural network [12, 13], etc. The basic feature extraction of these methods is complex, real-time is poor, and accuracy is low under field conditions. However, through the input of large-scale image data and iterative training, deep learning can independently extract the key features of the target, which has strong adaptability and robustness [14, 15].

With the emergence of the YOLO (You Only Look One) algorithm, this algorithm can make the computer identify items with higher accuracy. Therefore, many studies are trying to improve the YOLO algorithm, so that the algorithm has a higher precision or a lighter weight model for a certain agricultural product. In 2019, Yunong Tian et al. proposed an improvement to YOLOv3 for apple recognition in different stages of growth. The author added the DenseNet method to YOLOv3 to deal with the low resolution feature layer in the network, thus improving the propagation of the feature and improving the performance of the network [16]. In 2021, Yanchao Zhang et al. improved YOLOX for the recognition of Holly fruit and added the FocusNet to the backbone network, which improved the detection performance of YOLOX for Holly fruit [17]. There have also been some studies on green walnut in recent years. Jianjun Hao et al. in 2022 improved the YOLOv3 model for the identification of walnut. The author introduced the MobilNet-v3 backbone network to replace the YOLOv3 backbone network to improve the detection ability and lightweight of the model [18]. In 2022, Kaixuan Cui et al. improved the YOLOv4 for the recognition of green walnut. The author used the MobilNet-v3 as the backbone feature extractor for lightweight YOLOv4 [19]. However, all of these studies improved in the previous series of the YOLO. The YOLOv7 appeared in 2022. Compared with the previous series, the YOLOv7 has been greatly improved in terms of detection speed, detection time, and detection accuracy. Therefore, this study improved the YOLOv7 to find a more suitable detection model for green walnut.

In this study, a detection network based on YOLOv7 was designed. By adding an attention mechanism to YOLOv7, the detection accuracy of YOLOv7 against walnut was improved, which was used for real-time detection and counting of walnut fruit in complex scenes. The main contributions of the paper are summarized as follows:

(1) A novel image dataset for detecting walnut on trees are constructed, which containing 550 images and corresponding annotations. There are more than 3000 walnut objects in the dataset, which can be used for further research in the field.
(2) A novel walnut detection model named YOLOv7-TRS is proposed, which integrating three network modules, i.e., Transformer, ResNet, SimAM, in YOLOv7 framework, to enhance the feature representative ability.
(3) Extensive experiments are conducted to confirm the effectiveness of the proposed YOLOv7-TRS model in detecting green walnuts, comparing with existing object

detection models, which could potentially benefit farmers and other stakeholders in the walnut industry.

2 Data

2.1 Data Acquisition

The green walnut data was collected in Lantian, Xi'an, China. The data we collected in walnut orchards during sunny weather conditions from 11 am to 7 pm. A total of 550 images were collected. And the study collected two groups of green walnut image in a ratio of 7:5. The image format is jpg. Then, enlarge the first group of image data (there are 350 images in the data) from 350 to 7350 using data augmentation methods to generate the dataset for training. And the other images were used for testing and detecting.

2.2 Data Annotation

This study uses the LabelMe annotation tool to annotate the green walnut images which we collected, and to obtain the annotation file in JSON (JavaScript Object Notation). LabelMe is an image annotation tool developed by MIT's CSAIL (Computer Science and Intelligence Laboratory) that allows us to create customized tagging tasks or perform image tagging.

After annotating the data, convert the standard file in JSON to XML (Extensible Markup Language) and TXT format for easy follow-up under the different models.

2.3 Data Augmentation

The Convolutional Neural Networks often need many training samples to effectively extract the feature of images to output recognition results. To enhance the richness and the size of the image database, the study amplified the image using methods in the PIL (Python Image Library). PIL is the most commonly used image-processing library in Python. PIL supports image storage, display, and processing. And it can handle almost all image formats. The study used PIT to process the dataset in various forms. Such as flip, flipud, cutout, Gaussian blur, and affine. Subsequently, the dataset was expanded as shown in Fig. 1 (the study uses the dataset in the second set).

After augmentation of the image data, will have some images which used the cutout data augmentation have some labels in the clipped section. In order to prevent the influence of the experimental results, a relabel is carried out, and the faulty labels are deleted.

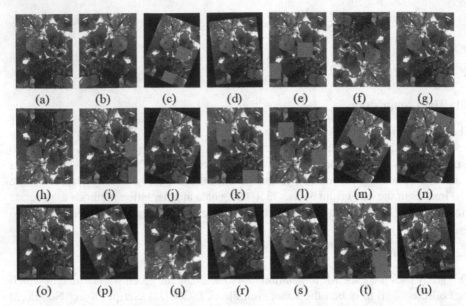

Fig. 1. Image augmentation methods: (a) original image, (b) Fliplr and GaussianBlur, (c) Affine and Cutout, (d) Affine, (e) Cutout, (f) Flipud, (g) GaussianBlur, (h) Fliplr and Flipud, (i) Cutout and GaussianBlur, (j) Affine, (k) Cutout, (l) Cutout and Flipud, (m) Cutout and Affine, (n) Cutout and Affine, (o) Affine, (p) Affine, (q) Flipud and GaussianBlur, (r) Affine, (s) Affine and GaussianBlur, (t) Cutout, and (u) Flipud and Affine.

3   Methods

3.1   YOLOv7

The YOLO [20] series is the first work of a one-stage object detection algorithm that has shown remarkable performance in the detection of agricultural products. The idea of YOLO series is to divide the target image into a grid shape, and calculate the position of the target object by obtaining the horizontal coordinate (x) and vertical coordinate (y) of the geometric centre of the target object and the width (w) and height (h) of the target object. The YOLO model is divided into two stages: training and prediction. The tasks of the two stages are different:

(1) Training phase. During the training phase, if the object centre falls into the grid, then the grid is labelled with the object's label (including xywh and category). This is how YOLO sets its training label.
(2) Test phase. Since YOLO had been taught during the training to predict which objects the centre would fall on the grid, Yolo naturally did the same.

The YOLOv7 [21] is one of the newest models in the YOLO collection. YOLOv7 improved the network architecture, model parameterization, and dynamic label assignment technology compared with the previous series. This improvement makes the YOLOv7 faster and more accurate than previous models. The YOLOv7 is divided into two training methods with and without an auxiliary head. This experiment uses the training without an auxiliary head. The Loss function used by YOLOv7 is GIoU Loss. Set the union area of the predicted target boundary box and the real target boundary box as U, and the intersection area of the predicted target boundary box and the real target boundary box as I. Set the predicted target boundary box and the real target boundary box as a minimum rectangle, and set the rectangle as A, then the specific formula of the loss function is as follows:

$$L_{GIoU} = 1 - \left(\frac{1}{U} - \frac{A-U}{A} \right) \tag{1}$$

3.2 Transformer

The Transformer [22] is the fourth major deep learning model after MLP, CNN, and RNN. Nowadays, the Transformer model is widely used in natural language processing, recommendation systems, computer vision, and speech. Compared with traditional models, the Transformer model has the characteristics of high parallel efficiency, low information transmission loss, and high information fusion rate. Transformer is essentially an Encoder-Decoder architecture. Therefore, the Transformer in the middle can be divided into two parts: the encoding component and the decoding component. The encoding component is composed of the multi-layer encoder. The decoding component is also composed of decoders with the same number of layers. Each encoder consists of two sub-layers: the Self-Attention layer and the Position-wise Feed Forward Network. The structure of each encoder is the same, but they use different weight parameters. Input from the encoder flows first into the Self-Attention layer. It allows the encoder to use information from other words in the input sentence when encoding a particular word. Decoders also have these two layers in encoders, but there is also an encoded-decoder Attention layer between them, which helps the Decoder focus on the relevant part of the input sentence. In the YOLOv7-TRS, we used the transformer to replace some of the convolutional blocks in the head of the YOLOv7 network.

3.3 ResNet

The ResNet [23] is a classical neural network that serves as the backbone of many computer vision tasks. Before ResNet was proposed, all neural networks were composed by the superposition of a convolutional layer and the pooling layer. It is believed that the more layers of convolution layer and pooling layer, the more complete image feature information can be obtained, and the better the learning effect will be. However, in the actual test, it is found that with the superposition of the convolution layer and the pooling layer, the learning effect is not getting better and better, but two problems: 1. Gradient vanishing and gradient explosion; 2. Degradation problem. In ResNet, the paper uses Batch BN (Normalization) layer and data preprocessing to solve the problem of gradient disappearance and gradient explosion. In order to solve the degradation problem in the deep network, ResNet artificially makes some layers of the neural network skip the connection of the next layer of neurons and connect the other layers, weakening the strong connection between each layer. Such neural networks are called ResNets. The paper of ResNet proposes a residual structure to reduce the degradation problem, and with the continuous deepening of the network, the effect is not worse, but better. The core idea of ResNet is to introduce an identity shortcut connection to skip one or more layers. The network improves network performance by simply stacking identity mappings on the current network. In the YOLOv7-TRS, we add the ResNet network to the backbone. Compared to other networks, ResNet achieved the best performance.

3.4 SimAM

The SimAM [24] is a Simple Parameter-Free Attention Module for Convolutional Neural Networks. SimAM is proposed to solve two problems with existing attentional infrastructure modules:

(1) The existing building blocks of attention can only refine features in one of the channel or spatial dimensions, but lack flexibility in Spaces where both space and channel vary simultaneously.
(2) The existing building blocks of attention structure often needs to be based on a series of complex operations, such as pooling.

This attention is based on well-known neuroscience theories that suggest the importance of optimizing energy functions to harness neurons. The SimAM proposed optimizing the energy function to capitalize on the importance of each neuron and derived a fast analytic solution to the energy function in less than 10 lines of code. Compared with other attention mechanisms, SimAM achieved the best performance.

3.5 YOLOv7-TRS

The weight capacity of the original YOLOv7 has a better detection effect. However, the detection effect has some space for growth. As a build-up on YOLOv7, the proposed YOLOv7-TRS added some modules to further increase the detection accuracy of the model. The network structure network of YOLOv7-TRS is show in Fig. 2. And the component of YOLOv7-TRS is show in Fig. 3. In the network, the backbone is used to extract features from image which was input. Traditionally, the more layers of the network, the richer the abstract features of different levels can be extracted in the CNN network. And the deeper the network, the more abstract the features extracted, the more semantic information. But in fact, simply increasing the depth of the network can easily lead to gradient disappearance and gradient explosion. In order to avoid gradient disappearance and gradient explosion caused by increasing the number of network layers, ResNet network is added to ELAN module in backbone. By increasing the number of network layers, the effect of image feature extraction can be improved. The improve of ELAN is show by ELAN-R in Fig. 3.

In This study, CNN is used to extract the features in the image, and then these features are input into the Transformer model as an input sequence, so as to realize the detection of walnut in the walnut image. The attention mechanism in the Transformer model adaptively focuses on information at different locations in the sequence, which makes the model better able to capture key information of objects at different scales,

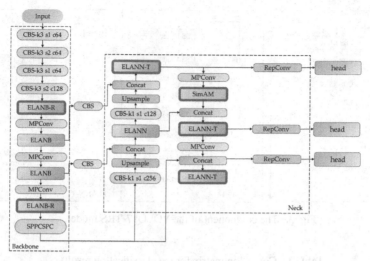

Fig. 2. The network structure of YOLOv7-TRS.

and better able to handle relationships and semantic information between objects, thus improving detection accuracy. With better generalization capabilities, Transformer can adapt to new scenarios and data to improve model reliability and robustness. Therefore, in order to improve the expression ability, generalization ability and interpretability of the feature representation of the model, as well as better scalability, we add Transformer module to the neck. The improve of ELAN is shown by ELAN-T in Fig. 3.

Finally, the SimAM module uses similarity measures to learn relationships between features, thereby increasing semantic similarity in feature representations while reducing redundant information between features. This helps to improve the model's ability to understand and express the input data, thus improving the model's performance and generalization ability. Adding the SimAM module to neck can improve the feature representation capability of the model, reduce the feature redundancy information, improve the generalization ability and interpretability of the model, and have lower computational complexity. These characteristics make SimAM an effective neck module that can be applied to various neural network models to improve model performance and application value.

4 Experiments

4.1 Performance Evaluation Metrics

The study samples were divided into four types: true positive (TP), false positive (FP), true negative (TN), and false negative (FN). The combinations matrix for the classification results is shown in Table 1.

The study used Precision (P), Recall (R), F-Measure (F1) and mean Average Precision (mAP) as the performance evaluation metrics. They are defined as follows:

$$P = \frac{TP}{TP + FP} \times 100\% \tag{2}$$

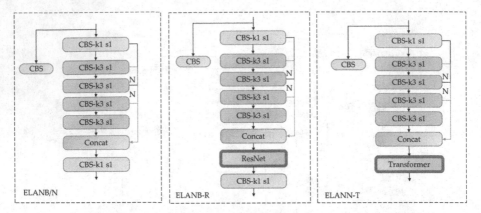

Fig. 3. The component of the YOLOv7-TRS model.

Table 1. Confusion matrix for the classification results.

Detect result	Positive example	Negative example
Positive example	TP	FN
Negative example	FP	TN

$$R = \frac{TP}{TP + FN} \times 100\% \tag{3}$$

$$F_1 = \frac{2 \times P \times R}{P + R} \times 100\% \tag{4}$$

$$mAP = \frac{1}{n} \sum_{i=1}^{n} \int_0^1 P(R)dR \tag{5}$$

4.2 Experimental Setting

The study was implemented on the GPU of the NVIDIA 3090 card, using CUDA 11.3.0 and CUDNN 8.2.2 parallel programming under the Linux environment (Ubuntu 18.04 LTS operating system). Pytorch 1.11.0 machine learning computational framework, Python 3.8, and OpenCV 4.5.4 were also used to implement these models.

The YOLOv7 -TRS uses the weight of YOLOv7 for training. After each iteration, save and update to the latest weight file, which will be used as pre-weight when resuming training after training interruption. Save the weight file for each iteration as well. The input resolution of the three models was 640 × 640, and the number of iterations was set to 300. The procedure of experimental is shown in Fig. 4. When the walnuts were ripe, take the two groups of pictures of the green walnut in the orchard as the original image data. The first set had 350 images and the second set had 200 images.

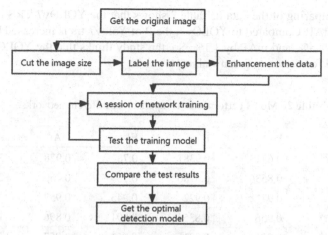

Fig. 4. Flow chart of build training model.

4.3 Results Comparison

The loss curve of the detection model is shown in Fig. 6. Faster R-CNN and RetinaNet approaches stable after 80 iterations, and YOLOv5s, YOLOv5x, YOLOv7, and Yolov7-TRS begin to stabilize after 20 iterations. As can be seen from the figure, Retina net, YOLOv5s, YOLOv5x, YOLOv7 and Yolov7-TRS have similar loss rates after stabilization, all lower than 0.1. According to the trend of the emergence curve, these models all learn the features of the object well, and the loss value after stability is less than 1, indicating that the models can be used for detection (Fig. 5).

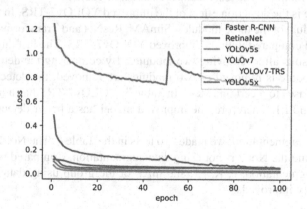

Fig. 5. The change in loss function.

The study used P, R, F1, and mAP as the accuracy index of the model in the result. And used the Faster R-CNN [25] model, the RetinaNet [26] model, the YOLOv5s model, the YOLOv5x model, the YOLOv7 and the YOLOv7-TRS model to train the dataset to get the accuracy index. Table 2 shows the model performance comparison of different networks. Comparing of the data in Table 2 shows that the YOLOv7-TRS has the best P, R, F1 and mAP. Compared to YOLOv7, the P of Yolov7-trs is increased by 2.1%, R by 1%, F1 by 1.5%, and mAP by 1.1%. So, the study thinks that the YOLOv7-TRS is the best model for detecting green walnut.

Table 2. Model performance comparison of different networks.

Accuracy index	P	R	F1	mAP	t (ms)
Faster R-CNN	0.6713	0.9387	0.78	0.928	20
RetinaNet	0.8536	0.6786	0.76	0.759	26
YOLOv5s	0.97	0.922	0.945	0.957	16.7
YOLOv5x	0.956	0.835	0.891	0.896	17.6
YOLOv7	0.971	0.931	0.951	0.957	12.7
YOLOv7-TRS	**0.992**	**0.941**	**0.966**	**0.966**	**26.9**

From the detection result, the YOLOv7-TRS can detect more green walnut. But the confidence in the result is low. The detection results of different models are as follows:

4.4 Ablation Study

Aiming at the detection effect of the improved YOLOv7 model, this study disassembled the model and conducted 6 groups of ablation tests respectively to verify the effectiveness of the proposed method. The test results are shown in Table 3.

The first line is the detection effect of the improved YOLOv7-TRS. In the next three groups, one module of the three modules, SimAM, ResNet and Transformer, is removed respectively and compared with the improved YOLOv7-TRS. The last three groups are the detection results after removing two modules. By comparing the detection results, whether one module is removed or two modules are removed, the detection effect of the model decreases to a certain extent. In Table 3, YOLOv7-TRS has a high detection effect on P, R and F1. Therefore, the improved model has a better detection effect on green walnut.

For the data augmentation, we made two tests in the Table 4. The No. 1 is using data augmentation, and the No. 2 is not using data augmentation. Compared with the group without data augmentation, P, R, F1, and mAP of the group using data enhancement have been greatly improved.

4.5 Problematic Situations

In model detection, the following four problems appear:

(1) The leaves were mistaken for the green walnut. In the image, some of the walnut leaves are similar in shape, color, and size to walnuts, which causes the computer to misjudge the leaves as walnuts (Fig. 7).

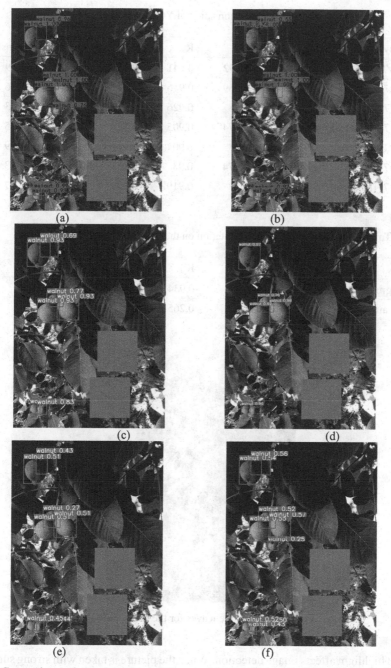

Fig. 6. Comparison of the detection of five models: (a) Faster R-CNN; (b) RetinaNet; (c) YOLOv5s; (d) YOLOv5x; (e) YOLOv7; (f) YOLOv7-TRS.

Table 3. Ablation study of YOLOv7-TRS.

Accuracy index	P	R	F1	mAP	t (ms)
YOLOv7-TRS	**0.992**	**0.941**	**0.966**	**0.966**	**26.9**
YOLOv7-Transformer-ResNet	0.969	0.916	0.942	0.936	27.6
YOLOv7-Transformer-SimAM	0.98	0.926	0.952	0.938	25.3
YOLOv7-ResNet-SimAM	0.974	0.905	0.938	0.929	20.5
YOLOv7-ResNet	0.98	0.907	0.942	0.926	20.9
YOLOv7-SimAM	0.974	0.93	0.951	0.942	20.3
YOLOv7-Transformer	0.98	0.919	0.949	0.934	22.7

Table 4. Influence of data augmentation on the performance of YOLOv7-TRS.

Test No.	P	R	F1	mAP
with augmentation	0.991	0.934	0.96	0.966
without augmentation	0.559	0.265	0.36	0.261

Fig. 7. Mistook the leaves for the green walnut.

(2) The highlight affects image detection. When the picture is taken with strong sunlight, the sun beam in the picture will block the walnut fruit and even make other walnut fruit more blurred, thus affecting the computer to identify the walnut in the picture (Fig. 8).

Fig. 8. The highlight affects image detection.

(3) Some green walnuts were not detected. In the walnut recognition process, some smaller or less obvious walnut fruits will be missed (Fig. 9).

Fig. 9. Some green walnuts were undetected.

5 Conclusions

Using artificial intelligence to accurately detect green walnut in natural environment is of great significance for the yield and quality estimation of walnut. In this paper, an improved detection model for green walnut based on YOLOv7 is proposed, and finally a detection model with strong detection ability for green walnut is obtained, which can better accurately detect green walnut in the natural environment. Based on the experimental results of this study, the following conclusions can be drawn:

(1) The complex background of natural environment has a great impact on detection. In the background of a complex natural environment, compared with Faster R-CNN and RetinaNet models, the loss function of YOLO series models for one-stage object detection converges faster and has better detection ability.
(2) In the YOLO series model, YOLOv7 has a stronger detection ability than YOLOv5 for green walnut. Although YOLOv7 has a high detection ability and can detect more green walnut, the confidence of YOLOv7 is low.
(3) After introducing the Transformer model, the ResNet network, and the SimAM attention into YOLOv7, YOLOv7-TRS is obtained. The P, R, F1 and mAP of the model can reach 99.1%, 93.4%, 96% and 96.6%. Compared with YOLOv7, it has a certain degree of improvement.

The method proposed in this paper solves the problems of low efficiency, low accuracy and large sample size of manual collection of green walnut in the natural environment, and provides technical support for the yield and quality estimation of green walnut. In future research, we will continue to expand the detection scope to include other kinds of crop detection and enhance the detection of crops by optimizing the detection network performance.

Acknowledgement. This work is supported in part by National Natural Science Foundation of China under grant No. 61902339, by the Natural Science Basic Research Plan in Shaanxi Province of China under grants No. 2021JM-418, by Yan'an Special Foundation for Science and Technology (2019-01, 2019-13).

References

1. Pei, D., Guo, B., Li, P., Zhang, Q., Xu, Y.: Investigation and analysis report of walnut market and industry. Agricult. Prod. Market 56–58 (2021)
2. Wang, C., Tang, Y., Zou, X., Luo, L., Chen, X.: Recognition and matching of clustered mature litchi fruits using binocular charge-coupled device (CCD) color cameras. Sensors **17**, 2564 (2017)
3. Fu, L., Tola, E., Al-Mallahi, A., Li, R., Cui, Y.: A novel image processing algorithm to separate linearly clustered kiwifruits. Biosyst. Eng. **183**, 184–195 (2019)
4. Reis, M.J.C.S., et al.: Automatic detection of bunches of grapes in natural environment from color images. J. Appl. Log. **10**, 285–290 (2012)
5. Cubero, S., Diago, M.P., Blasco, J., Tardáguila, J., Millán, B., Aleixos, N.: A new method for pedicel/peduncle detection and size assessment of grapevine berries and other fruits by image analysis. Biosyst. Eng. **117**, 62–72 (2014)

6. Wang, C., Lee, W.S., Zou, X., Choi, D., Gan, H., Diamond, J.: Detection and counting of immature green citrus fruit based on the local binary patterns (LBP) feature using illumination-normalized images. Precis. Agric. **19**, 1062–1083 (2018)
7. Nuske, S., Wilshusen, K., Achar, S., Yoder, L., Narasimhan, S., Singh, S.: Automated visual yield estimation in vineyards. J. FieldRobot. **31**, 837–860 (2014)
8. Borba, K.R., Oldoni, F., Monaretto, T., Colnago, L.A., Ferreira, M.D.: Selection of industrial tomatoes using TD-NMR data and computational classification methods. Microchem. J. **164**(4), 106048 (2021)
9. Fu, L., et al.: Banana detection based on color and texture features in the natural environment. Comput. Electron. Agric. **167**, 105057 (2019)
10. Goel, N., Sehgal, P.: Fuzzy classification of pre-harvest tomatoes for ripeness estimation – an approach based on automatic rule learning using decision tree. Appl. Soft Comput. **36**, 45–56 (2015)
11. Hamza, R., Chtourou, M.: Design of fuzzy inference system for apple ripeness estimation using gradient method. IET Image Proc. **14**(3), 561–569 (2020)
12. Kao, I.H., Hsu, Y.W., Yang, Y.Z., et al.: Determination of lycopersicon maturity using convolutional autoencoders. Sci. Hortic. **256**, 108538 (2019)
13. Mazen, F.M.A., Nashat, A.A.: Ripeness classification of bananas using an artificial neural network. Arab. J. Sci. Eng. **44**(8), 6901–6910 (2019)
14. Bai, X., Wang, X., Liu, X.L., et al.: Explainable deep learning for efficient and robust pattern recognition: a survey of recent developments. Pattern Recogn. **120**, 108102 (2021)
15. Muni Rathnam, S., Siva Koteswara Rao, G.: A novel deep learning architecture for image hiding. WSEAS Trans. Signal Process. **16**, 206–210 (2020)
16. Tian, Y., Yang, G., Wang, Z., Wang, H., Liang, Z.: Apple detection during different growth stages in orchards using the improved YOLO-V3 model. Comput. Electron. Agricult. **157**, 417–426 (2019)
17. Zhang, Y., Zhang, W., Yu, J., He, L., Chen, J., He, Y.: Complete and accurate holly fruits counting using YOLOX object detection. Comput. Electron. Agricult. **52–59**, 149–156 (2021)
18. Hao, J., Bing, Z., Yang, S., Yang, J., Sun, L.: Detection of green walnut by improved YOLOv3. Trans. Chin. Soc. Agricult. Eng. (Trans. CSAE) **38**(14), 183–190 (2022). https://doi.org/10.11975/j.issn.1002-6819.2022.14.021
19. Cui, K., Su, S., Cai, J., Chen, F.: Walnut ripeness detection based on coupling information and lightweight YOLOv4. Int. J. Circuits Syst. Signal Process. **16**, 239–247 (2022)
20. Redmon, J., Farhadi, A.: YOLO9000: better, faster, stronger. In: Proceedings of the 2017 IEEE Conference on Computer Vision and Pattern Recognition, Honolulu, 21–26 July 2017; pp. 6517–6525 (2017)
21. Alexey Bochkovskiy, C.-Y., Mark Liao, H.-Y.: YOLOv7: trainable bag-of-freebies sets new state-of the art for real-time object dectors. arXiv preprint arXiv:2207.02696v1 (2022)
22. Vaswani, A., Shazeer, N., et al.: Attention is all you need. arXiv preprint arXiv:1706.03762v5 (2017)
23. He, K., Zhang, X, Ren, S., Sun, J.: Deep residual learning for image recognition. arXiv preprint arXiv:1512.03385v1 (2015)
24. Yang, L., Zhang, R.-Y., Li, L., Xie, X.: SimAM: a simple, parameter-free attention model for convolutional netural networks. In: Proceedings of the 38th International Conference on Machine Learning, PMLR, vol. 139, pp. 11863–11874 (2021)
25. Ren, S., He, K., Girshick, R., Sun, J.: Faster R-CNN: towards real-time object detection with region proposal networks. arXiv preprint arXiv:1506.01497v3 (2016)
26. Lin, T.-Y., Goyal, P., Girshick, R., He, K., Dollar, P.: Focal loss for dense object detection. arXiv preprint arXiv:1708.02002v2 (2018)

Disease Recognition of Plants Leaves in Northern Shaanxi Based on Siamese Networks

Jinrong He[1](✉), Jiayi Zhang[1], Yani Sun[1], and Li Yang[2]

[1] College of Mathematics and Computer Science, Yan'an University, Yan'an 716000, China
hejinrong@yau.edu.cn
[2] School Hospital, Yan'an University, Yan'an 716000, China

Abstract. Crop disease image recognition is crucial for early detection and prevention of diseases in agriculture. With the development of deep learning technology, accurate identification can be achieved with large datasets. However, obtaining images of crop leaf diseases is challenging, and the use of Siamese networks can achieve good results with a small number of labeled samples, making it a suitable alternative to other deep learning algorithms. This paper explores the effectiveness of Siamese networks in crop disease image recognition of Northern Shaanxi area. We focuses on leaf diseases of three representative crops, including millet, apple, and jujube. By comparing Siamese network models based on ResNet34 and EfficientNetB0 with six traditional convolutional neural networks, such as AlexNet, VGGNet, GoogleNet, ResNet, DenseNet and MobileNet. The experimental results show that the Siamese network with ResNet34 as the backbone achieved highest accuracy on the millet leaf, apple leaf, and jujube leaf datasets, with accuracies of 99.2%, 99.8%, and 100% respectively. The results confirmed the effectiveness of the proposed method in improving disease recognition accuracies of crops in northern Shaanxi, which can have significant implications for agricultural production in the region.

Keywords: Siamese Networks · Leaf disease · Millet · Apple · Jujube

1 Introduction

Northern Shaanxi is one of the birthplaces of agriculture in China. The ideal harvest of crops in northern Shaanxi has greatly improved people's lives, but due to the influence of natural factors such as climatic conditions and farming patterns, pests, and weeds in crop production have shown a trend of increasing year by year. Crop diseases not only reduce the quality of crops, but also affect their quality and economic efficiency [1]. Due to the wide variety of pests and diseases, the identification is difficult, and the computer vision technology represented by deep learning overcomes the disadvantages of time-consuming, laborious and inaccurate identification of crop leaf diseases [2]. With the rapid development of deep learning technology, the accurate identification of crop diseases can be completed under large data sets [3]. Several common techniques for detecting and identifying crop diseases are support vector machines [4], decision

© ICST Institute for Computer Sciences, Social Informatics and Telecommunications Engineering 2024
Published by Springer Nature Switzerland AG 2024. All Rights Reserved
J. Li et al. (Eds.): 6GN 2023, LNICST 554, pp. 18–30, 2024.
https://doi.org/10.1007/978-3-031-53404-1_2

trees [6], and convolutional neural networks [7]. Thee Jun et al. [8] took five citrus disease images from the public dataset as research objects, experimented with the improved model R-ResNet, and compared it with VGGNet16, DenseNet121 and ResNet34. Experimental results show that the improved model R-ResNet accuracy is 3.9% higher than that of the ResNet34 model. Wang Linbai et al. [9] used five neural network models of AlexNet, VGG16, InceptionV3, ResNet50 and MobileNet for comparative analysis in the potato leaf disease identification experiment, and the InceptionV3 model had the best performance and an accuracy rate of 98% among the five models.

However, it is difficult to obtain images of crop leaf disease in the agricultural context, and the lack of training data will limit the scope of detection. Therefore, the Siamese network has a better ability to detect on small sample [10] datasets in crop disease identification tasks. Huo Shandong [11] uses the 34-layer residual network as the basic network structure to build a Siamese network for face recognition, and has achieved certain results. You Qingli et al. [12] used the Siamese-based algorithm to improve the identification accuracy of offline handwritten signatures by Hindi and Bangla by 1.68% and 6.73%, respectively, compared with the improved algorithm SigNet.

In this study, the Siamese network was designed to solve the problem of crop disease identification in northern Shaanxi, and the convolutional neural network model was designed using ResNet34 [13] and EfficientNetB0 [14] for the backbone network, and compared with seven other convolutional neural network models to verify the performance of the Siamese network. Experiments show that the Siamese network models based on ResNet34 and EfficientNetB0 outperform the classical models in the task of identifying crop leaf diseases, while the Siamese networks based on ResNet34 have a better effect, and the accuracy of the Siamese network on the disease image test data sets of millet [15], apple [16] and jujube [17] leaves reaches 99.2%, 99.8% and 100%, respectively.

The rest of the study is organized below. Section 2 describes related work. The third section introduces materials and methods for the identification of crop diseases in northern Shaanxi. Section 4 shows the results of the experiment. Finally, Sect. 5 presents the conclusions and outlook for the future.

2 Related Works

2.1 AlexNet

AlexNet won the 2012 ImageNet LSVRC with a much higher accuracy than the second place, bringing another high point in deep learning. The activation function of AlexNet, ReLU [18], is expressed as

$$F(x) = \max(0, x) \tag{1}$$

The experimental results show that the ReLU activation function is 6 times faster than the tanh [19] used by the previous neural network.

2.2 DenseNet

DenseNet uses a dense connectivity mechanism, which means that all layers are connected to each other, and each layer is connected to the previous layer in the channel (channel) dimension to achieve feature reuse as input to the next layer. This architecture has stronger gradient flow and mitigates the fact that gradient disappearance is more likely to occur when the network is deeper.

2.3 EfficientNet

EfficientNet uses a compound scaling means to scale the depth, width, and resolution of the network simultaneously to achieve a trade-off between accuracy and computational complexity FLOPS.

2.4 GoogleNet

GoogleNet was the first place in the 2014 ImageNet Challenge, and the four most important features in the model are the introduction of the Inception [20] structure (fusing feature information at different scales), the use of a 1×1 convolutional kernel for dimensionality reduction as well as mapping processing, the addition of two auxiliary classifiers to help with training, and the discarding of the fully-connected layer to use an average pooling layer (greatly reducing the model parameters).

2.5 MobileNet

The MobileNet model is a lightweight deep neural network proposed by Google for embedded devices such as mobile phones, and its core idea is depthwise separable convolution.

In depthwise (DW) convolution, only one kind of convolution kernel with dimension in_channels is used for feature extraction (no feature combination); in pointwise (PW) convolution, only one kind of convolution kernel with dimension in_channels 1*1 is used for feature combination.

Standard convolutional layer: Conv(F): $F \in R^{D_F \times D_F \times M} \longrightarrow G \in R^{D_G \times D_G \times N}$. Where DF denotes the spatial width and height of the square input feature map, and M is the number of input channels; DG denotes the spatial width and height of the square output feature map, and N is the number of output channels.

2.6 ResNet

The ResNet model won the 2015 ImageNet competition.The ResNet network is a modification of the VGG19 network.The changes are mainly reflected in the fact that ResNet directly uses the convolution of stride $= 2$ for downsampling and replaces the fully connected layer with the global average pool layer and adds the residual unit through the short-circuiting mechanism.The residual unit of the ResNet network is the residual unit of the VGG19 network. The residual unit of ResNet network.

There are two types of ResNet networks: shallow and deep networks.

The network architecture of ResNet34 can be seen that the residual edges of the ResNet34 network structure have solid and dashed lines respectively. The solid line structure has the same input and output dimensions; the dashed line structure has different input and output dimensions and uses the number of convolutional kernels to change the width, height, and depth of the feature map.

2.7 VGGNet

VGGNet won the runner-up in the 2014 ImageNet competition, VGGNet improved on AlexNet, the entire network used the same size of 3*3 convolution kernel size and 2*2 maximum pooling size. VGGNet uses a smaller 3*3 convolutional kernel and a deeper network. The stack of two 3*3 convolutional kernels is relative to the field of view of a 5*5 convolutional kernel, and the stack of three 3*3 convolutional kernels is equivalent to the field of view of a 7*7 convolutional kernel. This increases the CNN's ability to learn features.

3 Proposed Methods

3.1 Experimental Datasets

The experimental data set was manually collected images of crop diseases and pests in northern Shaanxi, including 441 images of three species of apple leaf diseases, 321 images of two species of jujube leaf diseases, and 232 images of two species of millet leaf diseases. The data were collected during July-September 2022, in the field under daytime with sufficient light, and taken with three cell phones with their own cameras. The iPhone 8 with 2GB running memory, main screen resolution of 1334 × 750, 7MP front and 12MP rear, and processor of Apple A11 + M11 coprocessor, respectively, was used with 4GB running memory, main screen resolution of 2688 × 1242 pixels, 7MP front and 12MP rear wide-angle and telephoto dual-lens cameras, and processor of The images were taken from 9:00 a.m. to 5:00 p.m., which allowed the data set to include different light intensities, angles, sizes, and sizes. This allowed the data set to include different light intensities, shooting angles, sizes, backgrounds, and pixels, ensuring the diversity of the data. The random sample is shown in Fig. 1.

The apple leaf disease data were collected from July to September 2022 at apple growing bases around Yan'an County, Shaanxi Province, with a total of 441 images of three diseases, including 117 images of spotted leaf drop, 101 images of brown spot, and 223 images of apple rust. The apple leaf spotted defoliation disease was collected in early July and mid-September 2022, the number of images was 117, mainly brown to dark brown round spots, surrounded by purple-brown halo, with clear edges, sometimes several spots fused and became irregular shape; apple leaf brown spot disease was collected in September 2022, the number of images was 101, the leaves turned yellow, the spots remained green around, forming Green halo; apple leaf rust was collected at the beginning of July 2022, the number of images is 223, when the first disease of the leaf, the front surface of the leaf appears shiny orange-red spots, gradually expand, forming a round orange-yellow spots, the edge of red. When the disease was severe, dozens of spots appeared on the leaves.

The data of jujube leaf disease were collected from jujube planting sites of farmers around Yanchuan County, Yan'an City, Shaanxi Province, with two diseases, totaling 321 spots, including 105 spots of jujube madness and 217 spots of jujube leaf mosaic disease. The foliar disease was collected with the diseased leaves as the background for direct photography. Jujube madness disease collection time for early July 2022, the number of images for 105, is a destructive disease of jujube, flowers into leaves, leaves small and yellow, in the form of clumps; jujube leaf foliar disease collection time for September 2022, the number of images for 107, leaves affected by the disease become small, distorted, deformed, leaves on the flesh of the leaf spot-like loss of green yellow, showing green-yellow, shades of foliar-like.

This collection of millet leaf disease data for farmers planted on the plains of Yan'an City, Shaanxi Province, Yanchang County, collecting a disease a pest, a total of 232, of which 107 millet white disease; 125 insect pests, is caused by the adult corn stem flea beetle. The number of images was 107, with yellowish-white stripes parallel to the veins on the adaxial surface of the leaf and grayish-white mold layer on the abaxial surface of the leaf when the air is moist; the number of images was 125, with white irregular longitudinal stripes on the epidermis.

Image augmentation techniques are used in the experiment include image contrast, random changes in brightness, and geometric transformations of images.

Fig. 1. (a) Millet infestation; (b) millet white disease; (c) apple brown spot; (d) apple rust; (e) apple spot defoliation disease; (f) jujube mosaic disease; (g) jujube madness

3.2 Experiment Setup

The experimental hardware environment is configured with the GPU of NVIDIA Ge Force GTX 3060 graphics card; the software environment is Windows1 system. The details of the hardware and software information in the experiment are shown in Table 1.

Table 1. Hardware and software information detail table.

Item	Type
CPU	11th Gen Intel(R) Core(TM) i7- 1165G7 @ 2.80 GHz 1.69 GHz
Memory	16 GB
GPU	NVIDIA Ge Force GTX 3060
OS	Windows10(64 位)

Since the data collected in this experiment were realized in natural environment with self camera, the size of the images were not consistent, which led to the irregular weights of the fully connected layer, and the network could not be trained, so the images were cropped to 224×224. The pre-trained [21] model was used in the code, and setting pretrained = True could ensure the reliability of the training results and accelerate the training speed.

The Siamese networks were evaluated on the data collected from three types of crops: millet, apple and jujube leaf diseases. The images in the training set are used to train and fit the model, and the images in the test set are used to evaluate the test accuracy of the model. Accuracy of the model. The parameters in the model are adjusted accordingly to ensure that the trained model has a higher test accuracy. The basic information of the dataset are shown in Table 2.

Table 2. Basic information about the dataset.

The dataset name	Number of categories	Total number of samples	The number of training set samples	The number of test set samples
Millet leaf diseases	2	3248	2275	976
Apple leaf disease	3	6174	4321	1853
jujube leaf disease	2	2968	2077	891

3.3 Siamese Network Model

The Siamese network model has M fully connected layers, each with Ns units, where h1,s represents the hidden vector in layer 1 of the first Siamese, while h2,s represents the second shared vector. We use specialized rectified ReLU units in the pre-M-1 layer, so for any layer s ∈ {1,...,M-1} eigenvectors are expressed by Formulas (2) and (3):

$$h_{1,m}\& = \max(0, W_{s-1,s}^{T} h_{1,(s-1)} + b_s) \tag{2}$$

$$h_{2,m}\& = \max(0, W^T_{s-1,s}h_{2,(s-1)} + b_s) \tag{3}$$

$W^T_{s-1,s}$ is the $N_{s-1} \times N_s$ shared weight matrix that connects the cells in layer s-1 to the cells in layer s and is the shared bias variable for layer s. After the (M-1)th feedforward layer, the features of each identical computation are compared by a fixed distance function for weighting the importance of the component distances, which defines the final Mth fully connected layer for connecting two identical networks. Smaller convolutional filters are used in Siamese networks to reduce the number of learning parameters while making the framework deeper (Fig. 2).

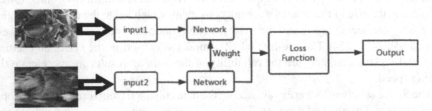

Fig. 2. Siamese network model

4 Proposed Methods

4.1 Millet Leaf Diseases

Table 3 compares millet leaf diseases between the two Siamese network models and 7 classical models, DenseNet and ResNet are the top two in the 7 classical models, and the accuracy of DenseNet training set and test set is 92% and 92.7%, respectively. The accuracy of ResNet training set and test set was 91.6% and 91.3%, respectively. The F1 value of DenseNet is 0.919 higher than that of ResNet by 0.005, so DenseNet is optimal among the 7 classical models. The accuracy of the Siamese network (EfficientNetB0) and the Siamese network (ResNet34) were higher than those of the 7 classical models, and the accuracy of the training and test sets of the Siamese network (EfficientNetB0) was 3.7% points and 4.5% points higher than that of the optimal models of the 7 classical models, respectively. 4.1% and 5.9% higher than the second-ranked ResNet model; The accuracy of the training and test sets of the Siamese network (ResNet34) was 6.5% points higher than that of the DenseNet model. 6.9% and 7.9% higher than the ResNet model. And the Siamese network (ResNet34) is more accurate than the Siamese network (EfficientNetB0).

Figure 3 shows the test results of millet leaf disease data on various models, from the figure, it can be seen that the Siamese network of ResNet34 as the backbone network gradually reaches convergence after the 5th iteration of training, and the loss value basically stabilizes below 0.1 and the accuracy is basically above 99% after the 10th iteration of training. The Siamese network of EfficientNetB0 as the backbone network has one oscillation at iteration 10, and then gradually decreases and reaches stability

Table 3. Comparative results on millet leaf diseases dataset

Model	Accuracy	Precision	Recall	F1-Measure	Training set accuracy (%)	Test set accuracy (%)
AlexNet	0.797	0.799	0.801	0.800	72.5	79.7
DenseNet	0.927	0.907	0.931	0.919	92	92.7
EfficientNet	0.898	0.899	0.902	0.900	90.5	89.8
GoogleNet	0.783	0.782	0.781	0.781	76.3	78.3
MobileNet	0.826	0.838	0.819	0.828	83.6	82.6
ResNet	0.913	0.913	0.915	0.914	91.6	91.3
VGGNet	0.710	0.713	0.713	0.713	69.2	71
Siamese Network (EfficientNet)	0.972	0.975	0.973	0.974	95.7	97.2
Siamese Network (ResNet)	0.992	0.995	0.998	0.996	98.5	99.2

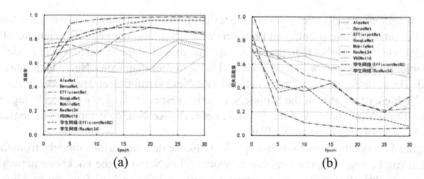

(a)   (b)

Fig. 3. Learning curves on millet leaf diseases dataset: (a) Accuracy curve; (b) Loss curve

after 20 iterations, and the overall loss value is higher than that of ResNet. Therefore, the millet leaf disease has a better performance on the Siamese network of ResNet34 as the backbone network.

4.2 Apple Leaf Diseases

Table 4 shows the comparison of Apple leaf disease in two Siamese network models and seven classical models, the accuracy of DenseNet and MobileNet is the same, but the F1 value of MobileNet is 0.005 higher than that of DenseNet. The accuracy of DenseNet training set and test set was 92.4% and 93.1%, respectively. The accuracy of MobileNet's training and test sets was 94% and 93.1%, respectively, so MobileNet was the best among the seven classical models. The accuracy of the Siamese network (EfficientNetB0) and

Table 4. Comparative results on apple leaf diseases dataset

Model	Accuracy	Precision	Recall	F1-Measure	Training set accuracy (%)	Test set accuracy (%)
AlexNet	0.840	0.850	0.831	0.840	82.4	84
DenseNet	0.931	0.940	0.915	0.927	92.4	93.1
EfficientNet	0.878	0.860	0.896	0.878	86.3	87.8
GoogleNet	0.870	0.889	0.855	0.872	88.1	87
MobileNet	0.931	0.960	0.906	0.932	94	93.1
ResNet	0.855	0.865	0.838	0.851	86	85.5
VGGNet	0.786	0.813	0.740	0.775	81.3	78.6
Siamese Network (EfficientNet)	0.962	0.968	0.996	0.982	98.5	96.2
Siamese Network (ResNet)	0.998	0.999	0.998	0.998	99.4	99.8

the Siamese network (ResNet34) were higher than those of the seven classical models, and the accuracy of the training and test sets of the Siamese network (EfficientNetB0) was 4.5% and 3.1% higher than that of the optimal model of the seven classical models, the MobileNet model, respectively. 6.1% and 3.1% higher than DenseNet, which ranked second; The accuracy of the Siamese network (ResNet34) training and test sets was 5.4% and 6.7% higher than that of the MobileNet model, respectively. 7.0% and 6.7% higher than the DenseNet model. And the Siamese network (ResNet34) is more accurate than the Siamese network (EfficientNetB0).

Figure 4 shows the test results of apple leaf disease data on various models, from the figure, it can be seen that the Siamese network of ResNet34 as the backbone network converges slightly faster than the Siamese network of EfficientNetB0 as the backbone network. ResNet34 gradually converges to zero loss value after 20 iterations, and the accuracy rate reaches more than 99% after 10 iterations.. In general, the recognition accuracy and convergence speed of ResNet34 and the Siamese network of EfficientNetB0 as the backbone network are better than the seven classical convolutional neural network models, while the recognition accuracy and model convergence stability of ResNet34 show better levels.

4.3 Jujube Leaf Diseases

Table 5 compares jujube leaf diseases between two Siamese network models and seven classical models, and the accuracy and F1 values of DenseNet are 0.036 and 0.033 higher than VGGNet, respectively. The accuracy of DenseNet training set and test set was 97.3% and 96.8%, respectively. The accuracy of VGGNet training set and test set is 94.2% and 93.2%, respectively, so DenseNet is the best among the seven classical models. The

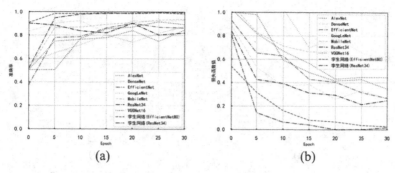

(a)   (b)

Fig. 4. Learning curves on apple leaf diseases dataset: (a) Accuracy curve; (b) Loss curve

Table 5. Comparative results on jujube leaf diseases dataset

Model	Accuracy	Precision	Recall	F1-Measure	Training set accuracy (%)	Test set accuracy (%)
AlexNet	0.898	0.916	0.896	0.906	90.3	89.8
DenseNet	0.968	0.969	0.969	0.969	97.3	96.8
EfficientNet	0.899	0.899	0.902	0.900	87.2	89.9
GoogleNet	0.932	0.938	0.932	0.935	92.0	93.2
MobileNet	0.937	0.945	0.936	0.940	92.3	93.7
ResNet	0.932	0.938	0.932	0.935	91.3	93.2
VGGNet	0.932	0.94	0.933	0.936	94.2	93.2
Siamese Network (EfficientNet)	1.000	1.000	1.000	1.000	98.5	100
Siamese Network (ResNet)	1.000	1.000	1.000	1.000	99.1	100

accuracy of the Siamese network (EfficientNetB0) and the Siamese network (ResNet34) were higher than those of the seven classical models, and the accuracy of the training and test sets of the Siamese network (EfficientNetB0) was 1.2% and 3.2% higher than that of the optimal model of the seven classical models, DenseNet model. 4.3% and 6.8% higher than the second-ranked VGGNet model; The accuracy of the training set and test set of the Siamese network (ResNet34) were 1.8% and 3.2% higher than that of the DenseNet model, respectively. 4.9% and 6.8% higher than VGGNet models. And the Siamese network (ResNet34) is more accurate than the Siamese network (EfficientNetB0).

Figure 5 shows the test result graphs of jujube leaf disease data on various models, from which it can be seen that the Siamese network of ResNet34 as the backbone network reduces the loss value to about 0.1 after the 15th iteration of training, and the accuracy rate is basically at 100% after 10 iterations. The Siamese network of EfficientNetB0 as

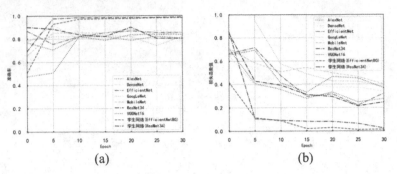

(a)   (b)

Fig. 5. Learning curves on jujube leaf diseases dataset: (a) Accuracy curve; (b) Loss curve

the backbone network has a loss value after 25 iterations The loss value is slightly lower than that of ResNet, and the accuracy is basically at 100% after 10 iterations. Overall, both ResNet34 and the Siamese network of EfficientNetB0 as the backbone network were better in apple leaf diseases, and ResNet34 was slightly better than EfficientNet.

5 Conclusions

In order to verify that the built Siamese networks have better recognition ability on three crops in northern Shaanxi, grain, apple and jujube, we compares the Siamese networks with ResNet34 and EfficientNetB0 as the backbone networks, and the classical convolutional neural networks among them, respectively. The experimental results have shown that, the accuracy of Siamese network with ResNet34 as the backbone network can achieve 98.5%, 99.4% and 99.1% for the training set and 99.2%, 99.8% and 100% for the test set on the millet leaf dataset, apple leaf dataset and jujube leaf dataset, respectively. The accuracy of the Siamese networks with EfficientNetB0 as the backbone network can achieve 95.7%, 98.5% and 98.5% for the training set on the millet leaf dataset, apple leaf dataset and jujube leaf dataset, respectively, and 97.2%, 96.2% and 100% for the test set, respectively. Siamese network with ResNet34 as backbone have achieved better performance.

In real life, the collection of crop leaf diseases is very difficult. There are not only seasonal restrictions when collecting crop leaf diseases, but also the selection of collection sites is difficult. Deep network model training relies on large-scale labeled samples, and when the number of samples is small, the performance of deep learning models is usually not satisfactory. Therefore, it is necessary to study the image recognition of small samples in crop disease recognition tasks.

Acknowledgement. This work is supported in part by National Natural Science Foundation of China under grant No. 61902339, by the Natural Science Basic Research Plan in Shaanxi Province of China under grants No. 2021JM-418, by Yan'an Special Foundation for Science and Technology (2019-01, 2019-13).

References

1. Zong, C.: Influence of green prevention and control technology on tobacco pests and insect pests and economic benefits (2021)
2. Su, T.: Research on Image Recognition of Crop Leaf Diseases Based on Transfer Learning. Shandong Agricultural University (2020). https://doi.org/10.27277/d.cnki.gsdnu.2020.000557
3. Yue, Y., Li, X., Zhao, H., Wang, H.: Image recognition of crop diseases based on improved VGG network. J. Agricult. Mechaniz. Res. **44**(06), 18–24 (2022). https://doi.org/10.13427/j.cnki.njyi.2022.06.003
4. Wu, Z., Zhao, J., Hu, X.: Plant disease detection based on feature fusion and SVM. Comput. Program. Skills Mainten. **2022**(02), 39–41 (2022). https://doi.org/10.16184/j.cnki.comprg.2022.02.052
5. Tian, Y., Li, T., Li, C., Park, Z., Sun, G., Wang, B.: Image recognition method of grape disease based on support vector machine. Trans. Chin. Soc. Agricult. Eng. **2007**(06), 175–180 (2007)
6. Jin, H., Song, K.: Application of decision tree algorithm in agricultural disease diagnosis. Contemp. Agricult. Machin. **202**(05), 76–77 (2007)
7. Sun, J., Tan, W., Mao, H., Wu, X., Chen, Y., Wang, L.: Identification of multiple plant leaf diseases based on improved convolutional neural network. Trans. Chin. Soc. Agricult. Eng. **33**(19), 209–215 (2017)
8. Tie, J., Luo, J., Zheng, L., Mo, H., Long, J.: J. South-Central Univ. Nat. (Nat. Sci. Edn.) **40**(06), 621–630 (2021)
9. Wang, L., Zhang, B, Yao, J., Yang, Z., Zhang, J., Fan, X.: Potato leaf disease identification and spot detection based on convolutional neural network. Chinese J. Agricult. Mechan. **42**(11), 122–129 (2021). https://doi.org/10.13733/j.jcam.issn.2095-5553.2021.11.19
10. Liu, Y., Lei, Y., Fan, J., Wang, F., Gong, Y., Tian, Q.: Review of image classification technology based on small sample learning. Acta Automat. Sinica **47**(02), 297–315 (2021). https://doi.org/10.16383/j.aas.c190720
11. Huo, S.: Face recognition and effect analysis using siamese network. Mod. Comput. **28**(20), 71–74, 83 (2022). https://doi.org/10.3969/j.issn.1007-1423.2022.20.015
12. You, Q., Li, G.: Offline handwritten signature authentication algorithm based on Siamese network. Comput. Appl. 1–5. http://kns.cnki.net/kcms/detail/51.1307.tp.20230320.1633.004.html. Accessed 21 Apr 2023
13. Heavy, Y., Lu, T., Du, Y., Shi, D.: Deepfake video detection method based on i_ResNet34 model and data enhancement. Comput. Sci. **48**(07), 77–85 (2021)
14. Chen, X., et al.: Application of EfficientNet-B0 and GRU-based deep learning on classifying the colposcopy diagnosis of precancerous cervical lesions. Cancer Medicine (2023)
15. Li, Y., Zhu, Y., Liu, P., Zhang, L.: Research on millet disease based on image recognition. Food Sci. Technol. Econ. **44**(03), 94–96 (2019). https://doi.org/10.16465/j.gste.cn431252ts.20190324
16. He, J., Shi, Y., Liu, B., He, D.: Research on external quality grading method of Fuji apple based on DXNet model. Trans. Chinese Soc. Agricult. Mach. **52**(07), 379–385 (2021)
17. Zhang, S., Huang, W.S., You, S.: Disease identification method of winter jujube based on Internet of Things and deep convolutional neural network. Zhejiang J. Agricult. Sci. **29**(11), 1868–1874 (2017)
18. Rubén, A., Fernando, H., Christoph, S.: ReLU neural network Galerkin BEM. J. Sci. Comput. **95**(2) (2023)
19. Liu, W., Meng, R., Qu, H., Liu, L.: Research on music genre recognition based on enhanced AlexNet. J. Intell. Syst. **15**(4), 8 (2020)

20. Kawakura, S., Shibasaki, R.: Suggestions of a deep learning based automatic text annotation system for agricultural sites using GoogLeNet inception and MS-COCO. J. Image Graph. **8**(4), 120–125 (2020)
21. Ji, X., et al.: Research and evaluation of the allosteric protein-specific force field based on a pre-training deep learning model. J. Chem. Inf. Model. (2023)

Application of MED-TET to Feature Extraction of Vibration Signals

Ningfeng Shan, Chao Jiang$^{(\boxtimes)}$, and Xuefeng Mao

ShangHai Dianji University, ShangHai 201306, China
jiangc@sdju.edu.cn

Abstract. Vibration signal reflects the operation status of the equipment. It is widely used in the field of mechanical fault diagnosis. However, the weak fault impact signal will be masked by the vibration noise, which makes it difficult to extract the fault features of the raw vibration signal. Aiming at this feature extraction problem, a hybrid method based on minimum entropy deconvolution (MED) and transient-extracting transform (TET) is proposed. First, the original signal is pre-processed by the MED method, which effectively reduces the interference of noise on the signal and enhances the impact component. Then, TET is used to extract the transient features of the pre-processed signal. Finally, the extracted transient information is used for fault diagnosis of rolling bearing. The validation of the method is carried out on simulated signals and Case Western Reserve University (CWRU) bearing data. Also, the proposed method is compared with other feature extraction methods. Those results show that the method can effectively extract the impact components in the vibration signal under strong background noise, which the effectiveness of the method is verified.

Keywords: Signal Processing · Minimum Entropy Deconvolution · Transient-Extracting Transform · Feature extraction · Vibration signal

1 Introduction

Rolling bearings are widely used in mechanical systems as important support parts of rotating machinery. The operating condition of bearings directly affects the overall performance of mechanical systems, so it is important to perform fault diagnosis on bearings. In recent decades, the early fault feature extraction technique for rolling bearings has attracted extensive attention. In the early fault stage of rolling bearings, due to the vibration signals that collected by sensors have no obvious features and are often submerged in strong background noise interference, therefor the basic problem of early bearing fault diagnosis is the extraction of shock pulses from the vibration signals which submerged by strong noise [1].

Considering that the bearing vibration signal is nonlinear and nonstationary, the time-frequency analysis (TFA) method can effectively characterize the information of the nonstationary signal in the time-frequency domain, therefore it is widely used in

J. Li et al. (Eds.): 6GN 2023, LNICST 554, pp. 31–42, 2024.
https://doi.org/10.1007/978-3-031-53404-1_3

bearing fault diagnosis. The traditional TFA methods mainly include short-time Fourier transform (STFT), wavelet transform (WT) and so on. However, these traditional time-frequency analysis methods have certain shortcomings. The STFT [2] has different criteria for selecting the length of the window function when dealing with different signals, which seriously affects the time-frequency resolution. The wavelet basis of wavelet transforms [3] is subject to human influence and cannot be changed once it is selected, which lacks self-adaptability. In order to enhance the ability to extract weak features of early bearing faults, Huang et al. [4] proposed empirical mode decomposition (EMD) in 1998, which decomposes the time series signal into the sum of several intrinsic mode functions (IMFs) and analyzes the IMFs of each order, which can enhance the transient features containing the fault frequencies. Shi et al. [5] proposed the Synchro squeezing transform (SST), which aims to obtain a clearer time-frequency representation and display the fault characteristics at high resolution. Yu et al. [6] proposed the Transient extracting transform (TET), which can effectively characterize and extract the transient components of the fault signal and obtain a more focused time-frequency energy representation. In summary, the time-frequency analysis method has made great progress in the pursuit of time-frequency resolution.

In order to overcome the influence of noise interference on early faint fault feature extraction, some nonlinear noise reduction methods such as wavelet noise reduction and morphological filtering [7] have been applied in bearing fault diagnosis. Although these methods can reduce the interference of noise on the signal, but they cannot enhance the shock component in the vibration signal. Minimum entropy deconvolution [8] (MED) can restore the spike impulse signal to the maximum extent by finding the optimal filter coefficients, so that the filtered signal has the minimum entropy value and the maximum cliff value, therefor the MED method can effectively enhance the early faint bearing fault signal. The literature [9] proposed to use the MED method to preprocess the vibration signal, which improves the cliff value of the original signal and enhances the impact characteristics of the fault, then combines the mathematical morphology method to extract the fault characteristics. In the literature [10], the time domain waveforms and envelope spectra before and after deconvolution by minimum entropy are compared and analyzed. It is concluded that the MED method can effectively improve the signal-to-noise ratio of vibration signals and bring out the periodic shock components of weak fault features that are submerged in strong noise.

In summary, for the extraction problem of weak transient features of bearing faults, this paper fully utilizes the signal enhancement characteristics of MED and the transient extraction characteristics of TET. A new time-frequency analysis method which based on MED and TET is proposed in terms of feature extraction. Firstly, the vibration signal is pre-processed by deconvolution, and the impact component of the early weak fault of the bearing is enhanced by MED to reduce the influence of noise; then the transient feature extraction operator (TEO) is used to extract the transient feature component. It is validated on the simulated signal and Case Western Reserve University bearing data set, and compared with the short-time Fourier transform, simultaneous compression transform, and wavelet transform for comparison.

2 MED-TET Algorithm

2.1 Minimum Entropy Deconvolution

The minimum entropy deconvolution algorithm [11] was first proposed by Wiggins and applied in processing seismic signals. MED aims to find an optimal set of inverse filter coefficients, that make the output signal after this inverse filter has the maximum cliff value and as close as the original impulse signal.

The basic principle of the MED algorithm is that the vibration signal is acquired by the sensor through the transfer path, which can be regarded as a convolution process with the structural filter. After that, the convolved output is mixed with the noise as the input of the inverse filter to output the reconstructed signal y(t). In order to achieve the desired effect, it is desired to be as similar as possible to the original input signal.

The output of the input after the structural filter, where represents the convolution operation:

$$s(t) = e(t)*h(t) \tag{1}$$

Inverse filter convolution expression of length L:

$$y(t) = \sum_{k=1}^{N} f(l)x(t-l) \tag{2}$$

The output after the inverse filter has the largest crag value, more structural impulse components and a more regular information distribution, thus achieve the minimum entropy value. When the bearing fails the impulse signal causes the system to resonate, the energy of the signal is concentrated toward the resonance band and the frequency components are more orderly, at this time the entropy value is smaller. Therefore, to find the minimum entropy value as the purpose of the knot deconvolution operation, can effectively highlight the impact component, so that the fault characteristics are more obvious.

The coefficient set of finding the inverse filter is achieved by the objective function method (OFM) for MED, expressed as:

$$O_4(f(l)) = \frac{\sum_{t=1}^{N} y(t)^4}{\left[\sum_{t=1}^{N} y(t)^2\right]^2} \tag{3}$$

The objective function (3) is derived as zero, at this time the signal has the maximum cliffness and the minimum entropy value, the expression is:

$$\frac{\partial O[f(l)]}{\partial f(l)} = 0 \tag{4}$$

Derivative for both sides of Eq. (2) yields:

$$\frac{\partial y(t)}{\partial f(l)} = x(t-l) \tag{5}$$

Combining Eq. (2) and Eq. (5), Eq. (4) can be expressed as:

$$b = Ag$$

where:

$$b = \left[\frac{\sum_{t=1}^{N} y^2(t)}{\sum_{t=1}^{N} y^4(t)} \right] \sum_{t=1}^{N} y^3(t)x(t-l)$$

$$A = \sum_{t=1}^{N} x(t-l)x(t-m)$$

$$g = \sum_{t=1}^{N} f(m) \tag{6}$$

b is the mutual correlation matrix of the input signal and output signal of the inverse filter. The inverse filter coefficients can be expressed by iterative calculation from Eq. (6) as:

$$g = A^{-1}b \tag{7}$$

2.2 Transient-Extracting Transform

This method is based on the Short Time Fourier Transform (STFT) for the Dirac function analysis to begin with. The STFT is expressed as:

$$G(t, \omega) = \int_{-\infty}^{+\infty} g(u-t) \cdot s(u) \cdot e^{-i\omega u} du \tag{8}$$

where $s(u)$ is the input signal and $g(u-t)$ is the moving window.

The Dirac function $\delta(t)$ is mathematically a shock function, which has zero values at all but zero, infinity at zero, and an integral over the definition domain. In the time domain, it appears in only one time position, so it is considered as an ideal model for representing signals with transient characteristics.

It is usually expressed as:

$$s_\delta(t) = A \cdot \delta(t - t_0) \tag{9}$$

where A is the amplitude of the impulse function and t_0 is the excitation moment. Due to the limitation of Heisenberg's inaccuracy principle, even for the impulsive signal with good time position, STFT makes the $\delta(t)$ energy representation smearing severely due to its own limitation. The derivation process of the group delay operator of the transient extraction transform is given below. Firstly, the short-time Fourier transform of this ideal signal model is performed:

$$G(t, \omega) = \int_{-\infty}^{+\infty} g(u-t) \cdot A \cdot \delta(t - t_0) \cdot e^{-i\omega u} du$$

$$= A \cdot g(t_0 - t)e^{-i\omega t_0} \tag{10}$$

The partial derivative of equation with respect to the frequency shift parameter is obtained as:

$$\partial_\omega G(t,\omega) = \partial_\omega \left(A \cdot g(t_0 - t) \cdot e^{-i\omega t_0} \right)$$

$$= -it_0 \cdot A \cdot g(t_0 - t) \cdot e^{-i\omega t_0}$$

$$= -it_0 \cdot G(t,\omega) \tag{11}$$

By deforming Eq., the excitation moment of the ideal signal model can be expressed as:

$$t_0(t, \omega) = i \cdot \frac{\partial_\omega G(t, \omega)}{G(t, \omega)} \tag{12}$$

For the ideal time-frequency analysis method, the energy should be concentrated only at the moment of excitation instead of being dispersed around. For this problem, a post-processing method called transient extraction operator is proposed, which is expressed as follows:

$$\text{TEO}(t,\omega) = \delta(t - t_0(t,\omega)) \tag{13}$$

It can be obtained as follows:

$$t_0(t,\omega) = \begin{cases} t_0, & t \in [t_0 - \Delta, t_0 + \Delta], \omega \in \mathrm{R}^+ \\ 0, & \text{otherwise} \end{cases} \tag{14}$$

Then it can be derived as follows:

$$\delta(t - t_0(t, \omega)) = \delta(t - t_0) \tag{15}$$

It can be found by the Eq. (15), only at the t_0 moment, the value of TEO is equal 1, at this time can be used to extract the moment of the time-frequency representation.

Given that Eq. (13) has transient extraction capability, the new time-frequency method is called transient extraction transformation, it is expressed as:

$$\text{TE}(t, \omega) = G(t,\omega) \cdot \text{TEO}(t,\omega) \tag{16}$$

To enable better diagnosis of faults, it is also necessary to recover the time domain signal of the signal to obtain more information. The known signal reconstruction equation for STFT is as follows:

$$s(t) = (2\pi g(0))^{-1} \cdot \int_{-\infty}^{+\infty} G(t,\omega) \cdot e^{i\omega t} d\omega \tag{17}$$

A similar reconstruction method can be used to extract the transient components from the results as follows:

$$s(t) = (2\pi g(0))^{-1} \cdot \int_{-\infty}^{+\infty} Te(t,\omega) \cdot e^{i\omega t} d\omega \tag{18}$$

2.3 The Method Proposed in this Paper

From the analysis in the previous two sections, it is known that TET has the ability to extract transient components compared with STFT, so that the shock components can have better time-frequency representation. MED has the ability to enhance the shock components and reduce the noise interference to the signal. Based on these two methods, this paper proposes a time-frequency analysis method for extracting early faint fault characteristics of bearings, and the flow chart of the algorithm is shown in Fig. 1. The method can be regarded as a combination of two algorithms, the MED algorithm can highlight the intensity of the weak impact components in the signal, and then the TET method uses the transient extraction operator to extract the individual transient components. After these two algorithms, the drowned weak shock signal can be successfully extracted and have a more focused time-frequency representation.

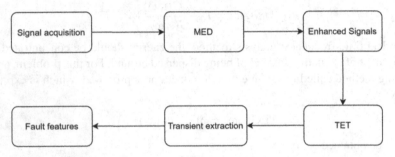

Fig. 1. Algorithm flowchart

3 Simulation Analysis

In order to verify the effectiveness of the proposed method for transient feature extraction, the simulated signal consists of a sinusoidal signal: $s = sin(2\pi \cdot 250t) \cdot e^{-120t}$, and the noise signal is randomly generated. The simulated bearing early weak fault shock signal is shown in Fig. 2(a), and the randomly generated Gaussian white noise is added. The signal after adding noise is shown in Fig. 2(b). It can be seen that the original weak shock component has been drowned by the noise.

The simulated signal is shown in Fig. 3. After MED to reduction the noise, not only the original strong impact component signal is enhanced, but also the weak impact component disturbed by noise is highlighted.

Then the transient extraction transformation is performed on the noise-reduced processed signal, and the transient characteristics of the shock components are extracted by the TEO operator pair. The time-frequency results of TET and MED-TET are given in Fig. 4. It can be seen that the proposed method not only successfully extracts the weak shock components from the simulated signal in the strong noise background, but also has a good energy concentration representation in the time-frequency diagram.

The transient classification of the two methods in Hilbert's envelope is given in Fig. 5. Where the characteristic frequencies of the shock signals and their high multiples can be clearly seen.

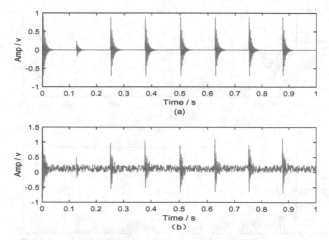

Fig. 2. (*a*) Simulate fault signal (b) Simulate fault signal with noise

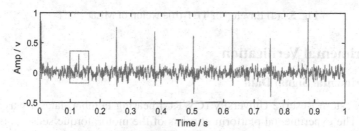

Fig. 3. Signal after reduction

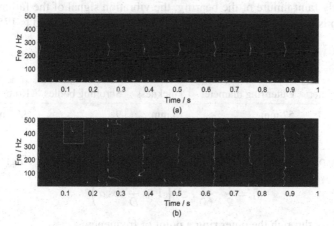

Fig. 4. Result of (a) TET (b) MED-TET

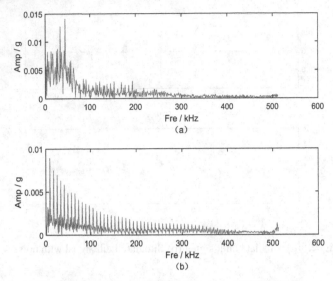

Fig. 5. (a) Envelop of TET (b)envelop of MED - TET

4   Experimental Verification

4.1   Experimental Signal Data

The Case Western Reserve University (CWRU) bearing number dataset was used for validation. The experimental platform consists of the motor, torque sensor, and power test meter. The model of drive end bearing is SKF6205, and the bearing data is shown in Table 1. The damage of the bearing is the single point loss of EDM. In order to reflect the early faint failure of the bearing, the vibration signal of the failure diameter of 0.1778 mm is selected. Sampling frequency is 12 kHz, motor speed is 1750 r/min.

Table 1. Parameters of SKF6205

Inner ring diameter	Outer ring diameter	Thickness	Scrolling bodies	Rotor Diameter
25 mm	52 mm	15 mm	9	7.94 mm

Frequency of the rolling body passing through a point of the inner ring:

$$f_i = \frac{1}{2} \cdot \frac{n}{60} \cdot z \cdot \left(1 + \frac{d}{D}\cos a\right) \tag{19}$$

Rolling body through the outer ring a point of frequency:

$$f_o = \frac{1}{2} \cdot \frac{n}{60} \cdot z \cdot \left(1 - \frac{d}{D}\cos a\right) \tag{20}$$

where n is the bearing inner ring speed, z is the number of rolling body, d is the ball diameter, D is the bearing raceway diameter, a is the bearing contact angle.

4.2 Fault Signal Analysis

The bearing fault signal and the results of several time-frequency analysis methods are shown in Fig. 6. From Fig. 6, it can be seen that the signal shows an obvious periodic shock signal, which is marked with a red box on the way, indicating that the bearing has failed. By doing the time-frequency analysis on it, the time-frequency display of STFT, SST and WT methods are given. But the energy of the fault signal is seriously dissipated after the short-time Fourier transform processing, which cannot be accurately located. wavelet transform and synchronous compression transform provide a very fuzzy time-frequency representation. While TET provides a clear time-frequency representation with a very concentrated energy, which is a feasible time-frequency analysis method.

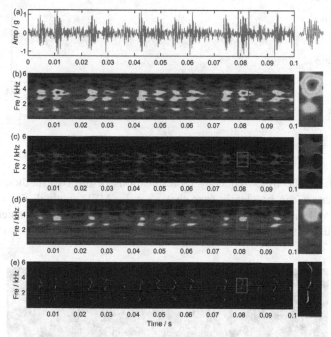

Fig. 6. (a) Fault signal (b) STFT (c) WT (d) SST (e) TET

At the same time, it was found that these time-frequency analysis methods did not accurately extract all the periodic signals, some micro-shock signals of the rolling bearing in the early stage are easily drowned in the noise. It can be seen that, the two places where the periodic signals should have appeared but it did not show up. Therefore, the method proposed in this paper will be used, using the minimum entropy deconvolution algorithm to pre-process the fault signal: highlighting those disturbed shock components and enhancing the fault characteristics.

The Fig. 7 shows the reconstructed signal of the fault signal after the minimum entropy deconvolution. In Fig. 7(b), it can be seen that the intensity of the shock signal after the noise reduction process is significantly higher than that of the original signal.

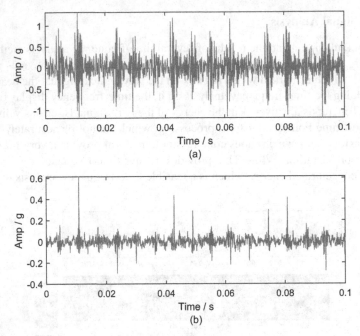

Fig. 7. (a) Fault signal (b) signal by MED

The results of the time-frequency representations of the two methods are given in Fig. 8, where it can be clearly seen that after the MED method, the weak faults are extracted and have a good time-frequency representation. The transient features are extracted using the operator for the original signal and the noise reduction signal

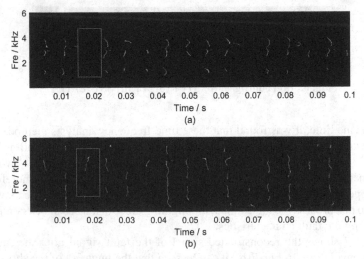

Fig. 8. (a) TET result (b)MED − TET result

respectively as shown in Fig. 9. In Fig. 9 the transient components of both signals are given and it can be seen that the signal after the MED method process, the transient extraction transformation can successfully extract the shock signal components and also extract the weak shock components that are drowned out by the noise. The envelope spectra of the transient components extracted by the two methods are given in Fig. 10.

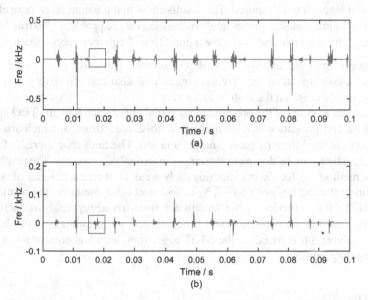

Fig. 9. Extracted transient components by (a) TET and (b) MED − TET

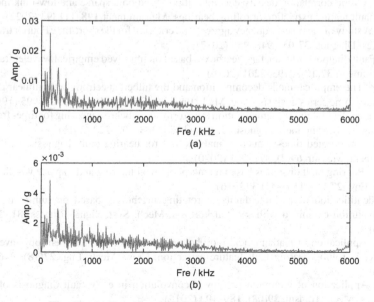

Fig. 10. (a) Envelop of TET (b) envelop of MED - TET

The MED-TET method clearly shows the characteristic frequencies of the rolling bearing faults in the plot, as well as the multiples of the fault frequencies.

5 Summary

In this paper, a transient extraction based on minimum entropy deconvolution of bearing fault vibration algorithm is proposed. The results show that the method is more effective than other traditional time-frequency analysis methods at the problem of extracting early faint bearing fault features, and also have a good time-frequency representation.

(1) Pre-processing of the fault signal by MED method effectively enhances the kurtosis and shock component of the original signal, and also can effectively reduces the interference of noise on the feature extraction.
(2) The transient features of the pre-processed signal are extracted by the TEO operator. The shock components which submerged in noise are extracted, which also have a more concentrated time-frequency representation. The fault characteristic frequencies can easily found by doing envelope spectrum analysis on them. The results show that the method is effective in extracting early weak fault characteristics of bearings.
(3) Comparing the method with STFT, WT, SST and other methods, the results show that MED-TET outperforms the traditional time-frequency analysis methods in terms of faint feature extraction and time-frequency representation. Furthermore, the TET algorithm is based on the STFT algorithm, which is comparable in terms of computing power.

References

1. Wang: Cyclic correlation density decomposition based on a sparse and low-rank model for weak fault feature extraction of rolling bearings. Measurement **198**, 111393 (2022)
2. Zhao. A study of the time-frequency aggregation criterion for the short-time Fourier transform. Vib. Test Diagnos. **37**(05), 948–956 (2017)
3. Zhu. Fault diagnosis of planetary gearboxes based on improved empirical wavelet transform. J. Instrument. **37**(10), 2193–2201 (2016)
4. Huang. The empirical mode decomposition and the hilbert spectrum for nonlinear and non-stationary time series analysis. Proc. Math. Phys. Eng. Sci. **454**(1971), 903–995 (1998)
5. Shi: Generalized stepwise demodulation transform and synchro squeezing for time–frequency analysis and bearing fault diagnosis. J. Sound Vib. **368**, 202–222 (2016)
6. Yu. A concentrated time–frequency analysis tool for bearing fault diagnosis. IEEE Trans. Instrument. Measur. **69**(2), 371–381 (2020)
7. Zhang. Bearing fault diagnosis based on morphological filtering and Laplace wavelet. China Mech. Eng. **27**(09), 1198–1203 (2016)
8. He. Identification of multiple faults in rotating machinery based on minimum entropy deconvolution combined with spectral kurtosis. Mech. Syst. Signal Process. **81**, 235–249 (2016)
9. Gong. Application of mathematical morphology method of minimum entropy inverse fold product in rolling bearing fault feature extraction. China Mech. Eng. **27**(18), 2467–2471 (2016)
10. Leng. Application of minimum entropy deconvolution in early fault diagnosis of rolling bearings. Mech. Transm. **39**(08), 189–192 (2015)
11. Wiggins. Minimum entropy deconvolution. Geoexploration **16**(1–2), 21–35 (1978)

A Novel Phase Congruency-Based Image Matching Method for Heterogenous Images

Donghua Li[✉] and Minzheng Li

School of Electronic Information, Shanghai Dianji University, Shanghai, China
superdong2016@outlook.com, limz@sdju.edu.cn

Abstract. Since different imaging principles of different ges lead to imaging results exhibiting nonlinear intensity differences, this phenomenon makes the traditional image alignment methods based on image gradients challenging. To solve this problem, instead of using image pixel intensity as the basis for image features, this paper searches for the relationship between phase and contour structure in digital images and summarizes this relationship as a new method for describing image structure. The method is light and contrast invariant, and furthermore, after phase voting, the method has a more powerful description between structure spaces. The proposed method is effective for image alignment of images from different sources.

Keywords: template matching · heterogenous image · phase congruency · phase-voting filter · Similarity strategy

1 Introduction

With the fast development of computer vision technology, in various industries, image processing for tasks that originally relied on manual matching, such as fault detection, medical imaging, satellite remote sensing, etc., is now beginning to be intelligently processed using computers vision. In different work scenarios, the images that need to be matched may be acquired by different devices. For homogenous images, the noise, rotation, and scaling problems arising from image acquisition can continue to be improved to some extent by algorithms, which provide the convenience for the design of subsequent algorithms. However, for heterogenous images acquired by different imaging principles (visible, CT, radar, infrared rays, etc.), even for the same object in the same scene, there are nonlinear differences and drastic amplitude variations between the two-dimensional images rendered under different imaging generation principles.

For current image matching methods, they can be broadly classified into two categories: Magnitude-based image matching and featured-based matching methods. In feature point-based papers [1–3], feature points are considered to be reliably detected

---

This Work Was Supported By the Shanghai Science and Technology Commission Science and Technology Plan Project: 21010501000.

J. Li et al. (Eds.): 6GN 2023, LNICST 554, pp. 43–51, 2024.
https://doi.org/10.1007/978-3-031-53404-1_4

and described mathematically between two images. In feature-point based algorithms, local invariant features are widely used for image alignment and scale invariant feature points (SIFT) [4] have been shown to perform well in many tests [5], because SIFT feature points are rotation and scale invariant for images, many researchers have proposed new descriptors based on them such as: oriented FAST, rotated BRIEF and other methods to improve the computational efficiency. However, they still do not escape from the effects of magnitude unreliability. These methods are sensitive to nonlinear radiometric differences [6] and produce poor alignment in the case of multimodal images (SAR, visible), and in the case of drastic changes in illumination, moreover, the phenomenon of Intensity inversion (see Fig. 1) leads to erroneous matching.

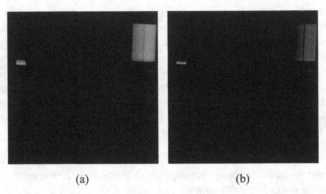

(a)   (b)

Fig. 1. The same scene is displayed under different imaging: (a) for visible light (b) for SAR, and it can be clearly seen that the two show inconsistencies in pixel gradients in selected areas.

The magnitude-based matching method does not perform feature detection compared to feature points though-such feature points are often Low repeatability in heterogeneous source images. Instead, a window of pre-pixel comparison is used to estimate the degree of similarity between the two images. However, the similarity measures normalized cross correlation (NCC), the sum of squared differences (SSD), and mutual information (MI) in template matching are based on linear, monotonic mapping rules and face the problem of unreliable magnitudes.

It is because of the recognition of the unreliability of traditional metrics based on magnitude and gradient that structural and shape features have started to be used as new similarity measures for image matching in recent related studies [7, 8]. As we can see in Fig. 1, although there are large differences in the intensity features of the images presented by different imaging principles, their structures and geometries are very similar. Obviously, because of the existence of the special phenomenon of Intensity inversion, we no longer use the image gradient to calculate the structural features.

Phase coherence methods have been shown to be better for extracting structural features of images [9], and descriptors constructed from phase coherence can reflect structural features of images in the presence of independent intensity distributions of heterogenous images [10].

This paper extends the concept of image gradient and constructs directional relationships between the phases of pixels in each block, inspired by anisotropic filters, and obtains the final direction of each block through a filter voting mechanism to provide a multimodal metric for similarity calculation.

2 Methodology

In this section, this paper will present the step-by-step design of the method, and its method flowchart is shown in Fig. 3. In this paper, we will propose a structural feature descriptor based which inspired by shaped-base image matching on the phase information of digital images, and on the basis of this descriptor, we will design a fast-matching algorithm for it.

2.1 Importance of Phase and Its Model

In digital image processing, in retrospect, many previous studies have used various means for image contour extraction, but these methods are often focused on the basis of image magnitude. Such as SOBEL, CANNY, SURF, etc., and these descriptors are usually more sensitive to illumination change and nonlinear contrast. According to the experiments of Ye [10], it was revealed that the original information of the image can be restored even when the amplitude of two images were exchanged and their own phase is retained, indicating that the contour structure features of the image are determined by the phase of the image, and phase has superior performance than the amplitude. Naturally, we turned to use the phase information to extract features from the image.

In past studies of phase coherence, it was found that image features are detected at the phase maximum of the Fourier component a study [12] has shown that the model and human vision are highly consistent for structure extraction of images. For a signal, its Fourier component can be expressed as $f(x) = \sum_n A_n cos(\phi_n(x) - \overline{\phi}(x))$, where A_n denote the magnitude of the nth Fourier component and ϕ_n denotes the phase of this component at x. The model that defines the phase consistency is

$$pc(x) = max_{\varphi(x)\in[0,2\pi]}\frac{\sum_n A_n cos(\phi_n(x) - \overline{\phi}(x))}{\sum_n A_n(x)} \qquad (1)$$

The value of $\overline{\phi}(x)$ represents the amplitude-weighted average local phase angle of the Fourier term at X that maximizes the equation. By expanding all the Fourier components in the complex plane, the phase consistency is computed for each e^{ix}. . For two-dimensional digital images, we choose quadrature pair filters (LOG-GABOR filters) to obtain the phase and amplitude of each Fourier component in the real and imaginary parts. The expression of the filter in the frequency domain is

$$g(\omega) = exp\left(\frac{-(\log(\omega/\omega_0))^2}{2(\log(k/\omega_0))^2}\right) \qquad (2)$$

where ω_0 denotes the center frequency. In order to ensure that the filters are geometrically similar in different directions, the value of k/ω_0 is kept relatively constant as much as possible when the center frequency ω_0 is changed, and the change of this term leads to a change in the filter bandwidth. For the real and imaginary components to be calculated, the responses corresponding to the even-symmetric (cosine) and odd-symmetric (sine) responses of the Log-Gabor filter and, are expressed as follows

$$[e_n(x, y), o_n(x, y)] = [I(x, y)] * M_n^e, I(x, y) * M_n^o]$$ (3)

The corresponding amplitude A_n and phase ϕ_n can be calculated by (3)

$$A_n = \sqrt{e_n(x, y)^2 + o_n(x, y)^2}$$ (4)

$$\phi_n(x, y) = atan2(e_n(x, y), o_n(x, y))$$ (5)

Considering the effects of noise and blur on the filter, we improve the model of Eq. (1) by setting the filter radius threshold T on the numerator of Eq. (1), and T is the average noise response of the noise circle in the frequency domain. In order to avoid a too narrow distribution of the filter, which leads to a local approximation of the sinusoidal function and makes the phase consistency calculation almost constant to 1, we expand in the actual calculation in for the numerator, using a weighting function to sharpen the local features, is a sigmoid function. Finally, considering that the cosine function is not sensitive near the 0 frequency, we choose $\Delta\phi(x)$ to use as the model for the phase consistency calculation:

$$pc(x, y) = \frac{\sum_o \sum_n W(x, y)[A_n(x, y)\Delta\phi_n(x, y) - T]}{\sum_o \sum_n A_n(x, y) + \varepsilon}$$ (6)

$$\Delta\phi(x) = \cos(\phi_n(x) - \overline{\phi}(x)) - |\sin(\phi_n(x) - \overline{\phi}(x))|$$ (7)

Figure 2 shows the results of contour structure extraction using the model for different illumination, which we compare with the traditional amplitude-based approach:

Phase Voting Filter

The phase consistency model described above obtains the feature value of each pixel and does not build up spatial information with the surrounding pixels, which cannot effectively reflect the direction of feature changes. Although this model can effectively avoid the influence of pixel magnitude and accurately extract structural information, it is not sufficient to fully describe the feature distribution in local regions of the image. We must extend the new dimension and define a new metric between the phases of each pixel to construct the descriptors.

We borrow the concept of image gradient and consider a window template $W(x, y)$ selected on the image in the phase coherence model, partition the window into blocks of size N*N and calculate the gradient of the phase in the Cartesian coordinate system. The magnitude and direction of the phase gradient vector are calculated by the following equation:

$$\begin{bmatrix} g_m \\ g_\phi \end{bmatrix} = \begin{bmatrix} \sqrt{f_x^2 + f_y^2} \\ tan^{-1}\frac{f_y}{f_x} \end{bmatrix}$$

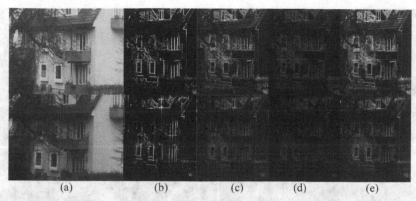

Fig. 2. (a) The original image under different illumination, (b) is the phase consistency model result, retaining the details of the window; (c) is the Prewitt edge result, detail contour is almost invisible; (d) is the Roberts edge result, detail is all gone; (e) is the Sobel result, retaining some details at the edge.

In order to exclude the interference of pixels with relatively small phase consistency in each block, we select the largest value of g as the main direction of each block, while for vectors with non-main directions, we use the direction voting mechanism to construct the following filter [11]:

$$filter = exp\left(-\left(\frac{r^2}{2\sigma^2}\right)\right)exp\left(1 - \frac{\pi/2}{|\pi/2 - |\alpha - \beta||}\right) \qquad (8)$$

where r is the relative distance between the location of the largest direction in each block and the vector of other locations, and σ is the scale of voting, which determines the effective voting interval. The first term of the filter determines that voters closer to the main direction will have a higher weight in the voting result. The second term of the filter indicates that the voter whose direction fits the main direction will cause more influence. The Fig. 3 shows the voting mechanism with the selected principal direction.

In each block, each pixel has a gradient amplitude and angle, which indicate the magnitude and direction of the pixel gradient change. In particular, if there are parallel but opposite angles in a local block, our filter does not care about the opposite direction, or rather, for the main direction, we need to exclude the gradient directions that are perpendicular to the main direction, because they do not change drastically enough to describe the variability between phases well, so our filter is more concerned with the parallelism between each pixel parallelism between them.

Similarity and Matching Strategy
In summary, we obtain the vector set of each BLOCK, and as mentioned before, after calculating the phase and the orientation of each BLOCK, our descriptor can powerfully capture the relationship between the internal structural features of the image and the structure of each pixel. Since this descriptor is independent of the intensity distribution pattern of the image, then for images with nonlinear but similar shapes, NCC is defined

Fig. 3. Filtering results in 45° direction

as a metric for image alignment, defined as

$$NCC = \frac{\sum_{k=1}^{n}(V_A(k) - \overline{V}_A)(V_B(k) - \overline{V}_B)}{\sqrt{\sum_{k=1}^{n}(V_A(k) - \overline{V}_A)^2 \sum_{k=1}^{n}(V_B(k) - \overline{V}_B)^2}} \tag{9}$$

where V_A, V_B is the consistency descriptor of phase PC_i in Figs. A and B, and the $\overline{V}_A, \overline{V}_B$ mean value of the descriptor corresponding to the image phase PC_i.

For a given block template orientation description $T(f_i) = [\theta_1, \theta_2, \ldots, \theta_n]$ and the orientation $O(f_i) = [\varphi_1, \varphi_2, \ldots, \varphi_n]$ in the image to be aligned, the similarity between them is calculated as in the following equation:

$$S(T(f_i), O(f_i)) = \sum_{i}^{i=n} \left| \frac{\pi}{2} - |\theta_i - \varphi_i| \right| \tag{10}$$

Under the constraint of the above equation, when the variation of the two directions is between $[-\pi, \pi]$, the proposed similarity will reach its maximum when the two directions are parallel, and all pixels parallel to the main direction can be extracted effectively. The parallelism is used to find the direction that best represents the largest change in phase gradient in each block after weighting, and this is used to constitute the spatial relationship of each phase in the block.

In the calculation of the block and the image to be aligned, we first calculate the average phase $\bar{v}$ in each block, and after the calculation of the Formula 10, the part of NCC less than a certain threshold is screened twice, and the directional similarity Formula 11 is used to discriminate again to improve the accuracy of the pairing.

3   Experimental Results

In the experiments designed in this paper, we selected 200 pairs of heterogenous images on which the designed method was evaluated. The same template size (150 pixels * 150 pixels) was randomly extracted from all SAR images and evaluated according to the proposed similarity method above. We choice the Correct Matching Rate (CMR) and the mean pixels error of correct matching to measure the accuracy of our methods. Where $CMR = SUM_{RIGHT}(I)/SUM(I)$, the average error rate represents the average Euclidean distance error at the pixel level between the region selected by the algorithm and the region where the template image is actually located in the correctly matched result. In Fig. 4, we show some matching results.

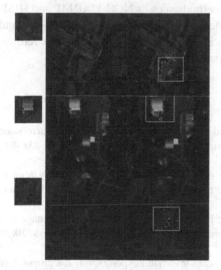

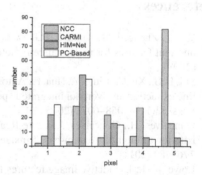

Fig. 4. Template matching result   **Fig. 5.** Error distribution of four methods

In Fig. 5, we compare the PC-based matching method with the other three methods. From the table, we can see that the PC-based proposed in this paper performs excellent, and the average error of matching is the lowest among the four methods, which is 2.11 pixels. Although the HIM-Net method can achieve a CMR of 78.8%, its average error pixel is 0.31 pixels points higher compared to our algorithm. The error distribution of the four methods is shown in the Fig. 5, from which we can see that the error of our algorithm is mostly distributed around one pixel point. In our experiments, the performance of our algorithm does not differ much from other algorithms when the image screen changes not drastically, but when faced with significant lighting changes or when the gradient information is unreliable, the accuracy of this algorithm reflects a greater advantage compared to other algorithms.

4 Conclusion

In this paper, we propose a method to turn to image phase as a basis for template matching when the gradient information is unreliable. The contours of 2D images are extracted by introducing the Log-Gabor variation in the communication field. The image from the image magnitude change caused by the illumination change is effectively avoided. In addition, the phase direction is calculated for each block using the voting filter mechanism, and the similarity measure is defined in the direction to ensure that the characteristics of the contour structure can be fully utilized. The proposed template matching method can precisely find the region corresponding to the template image in the heterogenous image, which provides a new phase-based solution to the image matching problem in the presence of nonlinear differences and intensity inversion between the two images and this method outperforms the methods such as NCC, CARMI, and HIM-Net. However, the problems targeted by the method in this paper are not extensive, and excellent results can be achieved in more specific environments. In future research, we will focus will be more on the problem of expanding the application area.

References

1. Yu, L., Zhang, D.R., Holden, E.J.: A fast and fully automatic registration approach based on point features for multi-source remotesensing images. Comput. Geosci. **34**(7), 838–848 (2008)
2. Sui, H.G., Xu, C., Liu, J.Y., Hua, F.: Automatic optical-to-SAR image registration by iterative line extraction and Voronoi integrated spectral point matching. IEEE Trans. Geosci. Remote Sens. **53**(11), 6058–6072 (2015)
3. Gonçalves, H., Gonçalves, J., Corte-Real, L.: HAIRIS: a method for automatic image registration through histogram-based image segmentation. IEEE Trans. Image Process. **20**(3), 776–789 (2011)
4. Lowe, D.G.: Distinctive image features from scale-invariant keypoints. Int. J. Comput. Vis. **60**(2), 91–110 (2004)
5. Mikolajczyk, K., Schmid, C.: A performance evaluation of local descriptors. IEEE Trans. Pattern Anal. Mach. Intell. **27**(10), 1615–1630 (2005)
6. Ye, Y., Shen, L.: Hopc: a novel similarity metric based on geometric structural properties for multi-modal remote sensing image matching. In: ISPRS Annals of Photogrammetry, Remote Sensing and Spatial Information Sciences; 12–19 July 2016, Prague, pp. 9–16. ISPRS, Gottingen (2016)
7. Li, Z., Mahapatra, D., Tielbeek, J.A.W., Stoker, J., van Vliet, L.J., Vos, F.M.: Image registration based on autocorrelation of local structure. IEEE Trans. Med. Imag. **35**(1), 63–75 (2016)
8. Rivaz, H., Karimaghaloo, Z., Collins, D.L.: Self-similarity weighted mutual information: a new nonrigid image registration metric. Med. Image Anal. **18**(2), 343–358 (2014)
9. Kovesi, P.: Image features from phase congruency, Videre. J. Comput. Vis. Res. **1**(3), 1–26 (1999)
10. Ye, Y., Shan, J., Bruzzone, L., Shen, L.: Robust registration of multimodal remote sensing images based on structural similarity. IEEE Trans. Geosci. Remote Sens. **55**(5), 2941–2958 (2017). https://doi.org/10.1109/TGRS.2017.2656380

11. Lu, J., Hu, M., Dong, J., Han, S., Su, A.: A novel dense descriptor based on structure tensor voting for multi-modal image matching. Chinese J. Aeronaut. **33**(9), 2408–2419 (2020). https://doi.org/10.1016/j.cja.2020.02.002. ISSN 1000-9361
12. Spacek, L.A.: The Detection of Contours and their Visual Motion. Ph.D. thesis, University of Essex at Colchester (1985)

Design of Image Based Optical Flow Tracking and Positioning System in Intelligent Assembly

Su Wang$^{(\boxtimes)}$ and Minzheng Li

School of Electronic Information, shanghai Dianji University, Shanghai, China
su.1165746435@qq.com

Abstract. The movement of workers in intelligent assisted assembly scenarios can cause model shaking, as well as issues such as obstruction, severe tilting, and scale changes of landmarks, which can lead to failed tracking of landmarks. Based on this, a high-precision and fast detection and tracking method is proposed. Firstly, the framework of the marker detection and tracking algorithm was established, and its working principle and process were explained; Secondly, coordinates are provided for subsequent algorithms through marker detection calculations; Then, the pyramid optical flow algorithm in the visual odometer is added to process the coordinates, and a tracking effect is added to the landmark coordinates to combat the problems caused by worker movement; Finally, when using the processed coordinates for pose solving, a nonlinear optimization matrix that combines the two-step degree method and Gaussian Newton method is added to enhance the algorithm's anti-interference ability. The simulation experiment results show that the marker detection and tracking algorithm runs quickly, with guaranteed tracking accuracy and efficiency. It can effectively solve related problems such as image jitter, occlusion, and scale changes, and effectively reduce the impact of lighting. The design has achieved the expected results.

Keywords: augmented reality · detection and tracking · pose solution · optical flow algorithm · nonlinear optimization

1 Introduction

With the rapid development of artificial intelligence, machine vision, and computer vision technology in recent years, the three-dimensional tracking and detection technology of visual SLAM (Simultaneous Localization And Mapping) in intelligent assisted assembly has developed rapidly in recent years, receiving in-depth research and widespread application.

Nister et al. [1] proposed the five point method for solving camera pose in the design of large-scale visual systems. Subsequently, Davis et al. [2] proposed a slow running monocular SLAM algorithm with significant bias in system pose estimation. In response to the problem of pose deviation for a single object, Bend et al. [3] proposed coordinate system transformation to solve motion parameters. Then Kaess et al. [4]

J. Li et al. (Eds.): 6GN 2023, LNICST 554, pp. 52–61, 2024.
https://doi.org/10.1007/978-3-031-53404-1_5

studied the performance and positioning accuracy of three feature extraction algorithms: SIFT (Scale invariant feature transform), SURF (Speed Up Robust Features), and ORB (Oriented FAST and Rotated BRIEF). The results showed that the ORB algorithm has the highest stability. Finally, Mur Artal et al. constructed a three threaded ORB_ SLAM [5] and proposed for monocular, binocular, and RGB D cameras based on ORB_ SLAM2 [6]. The visual positioning method proposed by Qi Naixin et al. [7] utilizes multi-scale and multi region extraction of ORB key points for optical flow tracking, which has good computational accuracy and speed. However, due to the fact that this algorithm is based on L-K optical flow tracking to estimate motion, it performs poorly when motion blur occurs.

Intelligent assisted assembly in industry is developed and applied based on the SLAM platform. Tang Jianjun et al. [8] designed an AR intelligent guidance system to assist in aircraft assembly, used to identify small and medium-sized rigid components in aircraft production sites, and combined it with dynamic AR assembly instructions in XML format to judge the assembly results. Westerfield et al. [9] designed an augmented reality motherboard assembly training system, which provides real-time feedback on warning information when training users. Liu Ran [10] designed an auxiliary assembly guidance system for key components of car doors, which estimates the pose of the assembly matrix and parts, and judges the assembly steps and assembly effects.

In actual intelligent assisted assembly scenarios, workers moving near assembly points in search of landmarks may experience motion blur, oblique occlusion, scale changes, and recognition and positioning failures due to the influence of reflective metal materials in the assembly. Therefore, this article based on the library platform of ORB_SLAM2,this article conducts research on detection and tracking algorithms for these issues include:

(1) Workers need to ensure both accuracy and tracking effect during actual assembly operations. Due to the fact that the positioning accuracy relies on the estimation of feature points in the image, and the tracking effect relies on real-time matching of each frame, a high precision ORB image fast detection and tracking algorithm framework is constructed for these two indicators in a low computational power intelligent assembly system;
(2) Assist workers in searching for landmarks before they can find them. During this process, it is necessary to overcome the issues of motion blur and image jitter, and the accuracy requirement is to be able to recognize the approximate position of the landmarks. To address the serious issues of motion blur and image jitter when using pure feature point matching, an optical flow algorithm is added to the algorithm flow framework;
(3) There will be a large amount of metal materials in the assembly workshop, which will cause metal reflection to affect the auxiliary assembly. Therefore, in order to avoid the sensitivity of ORB feature point extraction and classical L-K optical flow method to light as much as possible, the latest pyramid optical flow algorithm in the visual odometer was innovatively selected to assist in the processing of feature points.

2 Image Detection and Tracking Algorithm Framework

The entire detection and tracking algorithm framework is shown in Fig. 1. The algorithm loop consists of three parts, namely two input modules, two mode modules, and one pose processing module. Among them, the two input modules are video sequence frames and image markers pre-processed within the code, and the two mode modules are detection and tracking modules, respectively.

As shown in Fig. 2, when a video sequence frame is input, the first frame first enters the detection module, and the ORB algorithm extracts feature points and matches them with the already extracted Marker (marker) image. If no match is found, the current frame is skipped, and the next frame data is processed and re entered into the detection mode. If the match is found, the initialized tracking module is entered and the tracking mode is enabled, then enter the pose processing module, and the subsequent frames will directly enter the tracking mode for optical flow tracking. If the ROI area extracted by the mask [11] is too small or obstructed, Harris corners will be re extracted in the mask area [12]before optical flow tracking. If the tracking is lost, the current frame will be

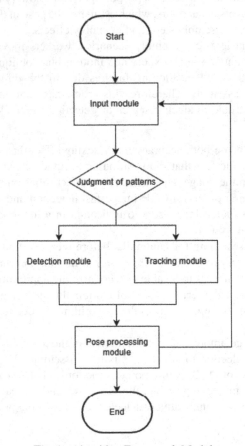

Fig. 1. Algorithm Framework Module

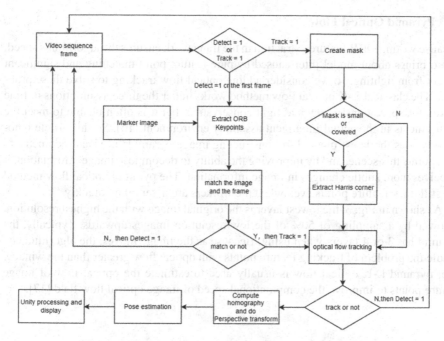

Fig. 2. Image detection and tracking algorithm framework

skipped, and the next frame data will be input for processing and enter the detection mode, If the tracking is successful, continue to maintain the tracking mode and then enter the pose processing module.

3 Algorithm Process Module Improvement

3.1 Marker Detection and Pose Determination

Firstly, using Fast feature points [13] to detect corner positions and combining with Brief descriptors to extract descriptive features [14]. After detecting the ORB feature points of the current frame and preset markers in the video, the BRIEF descriptor is matched using the Hamming distance that represents the similarity of the matching points. After traversing the feature points, find the distance between the two sets of points with the most similar dmin and the least similar dmax. Filter matching point pairs based on the minimum distance, retain coplanar matching point pairs with high similarity, and obtain homography transformation. Then filter out the matching point pairs with large errors again through RANSAC [15] (random sampling consistency). Finally, perform this matrix transformation on the four vertices of the landmark to obtain the coordinates in the current frame of the video and perform tracking processing using the optical flow algorithm.

3.2 Pyramid Optical Flow

Because writing ORB feature extraction in the framework ensures the algorithm's speed, it also brings about model jitter caused by pure feature point matching and significant impact from lighting, so we consider adding optical flow tracking to solve these problems. The classical L-K optical flow method works under the three assumptions of time continuity, space consistency and light conservation, but it is often unable to meet the requirements in the actual intelligent assembly environment [16]. So this article innovatively uses the latest method of constructing image pyramids in visual odometry to incorporate this scene, and by improving the ability to decompose image information, it perceives more subtle changes in image information. The pyramid optical flow method can still track feature points even when the workers are assembling quickly.

As shown in Fig. 3, the lowest layer is the original image with the highest resolution, followed by a sampling of 50% of the low resolution images upwards. Typically, the pyramid has 3 or 4 layers when using. Because as the image moves, the algorithm can handle the problem of tracking feature points with optical flow greater than the window size, pyramid L-K optical flow is usually used to estimate the optical flow of image feature points to improve the computational speed of image optical flow field [17].

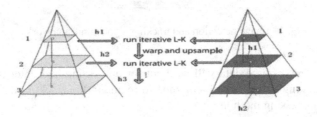

Fig. 3. Pyramid Optical Flow Algorithm

Therefore, the steps to use the L-K pyramid are: first, calculate the optical flow of the top-level image of the pyramid, calculate the initial value of the optical flow of the secondary top-level based on the results of the top-level optical flow, and then further estimate the accurate value of its optical flow. Finally, the initial value of the optical flow in the next layer is estimated using the calculated results of the optical flow in the next layer, and its exact value is calculated before continuing to the next layer until the bottom layer of the pyramid is reached [18]. Then, the obtained value is substituted into the p4p algorithm to solve its pose.

3.3 Pose Optimization

After detecting a frame of an image with landmarks using a camera calibrated on a chessboard, both C and x are known. Next, arbitrarily select a world coordinate system to solve the external parameter matrix Rt. Here, the world coordinate system is directly set on the landmark, and the world coordinate system is established at the center point of the landmark, with the z-axis component set to 0 and its positive direction facing

Fig. 4. Custom Coordinate System

outward to meet the right-hand coordinate system. Here, the coordinates of the lower right corner of the landmark can be defined as $(0.4, -0.3, 0)$.

Because the marker detection has obtained 4 pairs of matched, coplanar 3D points X and 2D points x, solving the camera's external parameter Rt matrix is a coplanar p4p problem. Here, the nonlinear optimization method is used to solve the optimization equation. The goal of optimization is to make the projection points of 3D points onto the 2D plane as close as possible to the 2D points, with the minimum error. Assuming there is a spatial point $x(x, y, z)$ and its corresponding point $p(u, v)$, the camera's internal parameter is C, and the external parameter is $R\&t$. The observation equation can be written as:

$$p = C(Rx + t) \tag{1}$$

Assuming the existence of N pairs of 3D-2D matching points, the overall error can be minimized by iterating $R\&t$ [19]:

$$minf(x) = \sum_{k=1}^{N} e_k(x_k, z_k)^T \Omega_k e_k(x_k, z_k) \tag{2}$$

Among them, e represents the vector error of the function. Ω is the inverse diagonal matrix of the covariance matrix, indicating the correlation between errors and the degree of attention. To solve this equation, an initial point x_{k0} and an iterative direction Δx are required, so the estimated value of the edge is $f_k(x_{k0} + \Delta x)$, and the error term becomes $e_k(x_{k0} + \Delta x)$. The second order Taylor expansion of the Error term using the idea of Gauss Newton method is as follows:

$$e_k(x_{k0} + \Delta x) \approx e_k(x_{k0}) + J(x_{k0})^T \Delta x + \frac{1}{2} \Delta x^T H(x_{k0}) \Delta x \tag{3}$$

$J(x_{k0})$ is the first derivative of e_k to x_k, in the form of Jacobian matrix, and $H(x_{k0})$ is the second derivative, called Hesse matrix. Due to the abandonment of the second-order degree when calculating the Hesse matrix in general, which can lead to a decrease in accuracy, the following calculations need to retain this second-order degree. At this point, the objective function is rewritten as:

$$f_k(x_{k0} + \Delta x) = c_k + 2b_k \Delta x + \Delta x^T H_k \Delta x \tag{4}$$

The term unrelated to Δx in the above equation is c_k, the coefficient of the first term is $2b_k$, and the coefficient of the second term is H_k. From the derivation, it can be inferred that c_k is the value of $f_k x_{k0}$, therefore Δf_k is:

$$\Delta f_k = 2b_k \Delta x + \Delta x^T H_k \Delta x \tag{5}$$

Therefore, according to the idea of the two step method, it is necessary to solve for the minimum change of Δf_k when Δx changes. So taking the derivative of this equation and making the derivative equal to 0 yields:

$$\frac{df_k}{d\Delta x} = 2b_k + 2H_b \Delta x \equiv H_k \Delta x = -b_k \tag{6}$$

Finally, the pose increment Δx of the optimization algorithm is solved and transformed into various camera coordinate systems. After pose optimization, the overall recognition accuracy of the algorithm has been improved by 2.7% under oblique occlusion and different lighting conditions.

4 Experimental Results

The hardware configuration environment used in the experiment is Intel(R) Pentium(R) CPU G4600,graphics card NVIDIA GeForce GTX 1050, operating system Windows10,software environment Microsoft Visual Studio 2017, Opencv3.4.15, and an external low-precision USB camera.

When comparing algorithms in this experiment, a low precision USB camera was selected to record a scene video with landmarks. The video was detected by capturing frames from different poses. If successful, four edges of the detected landmarks were drawn to visually demonstrate the effectiveness of the algorithm. When the error of the four edges drawn exceeds 50 mm, this data is actually detected as lost, so it is included in the recognition loss rate and not in the average error.

Table 1. Comparison of time consumption, recognition loss rate, and average error of each algorithm

Algorithm	SIFT	SURF	ORB	Gracker	Ours
Average time consumption(ms)	97.59	79.16	31.22	129.52	15.69
Identification loss rate(%)	18.19	9.71	18.32	15.85	7.62
Average error(mm)	17.37	10.69	17.26	16.82	9.87

In response to the situation where workers may experience tilting, occlusion, scale changes, and tracking recognition failure due to the influence of reflected light from metal materials when detecting markers during movement, the most classic algorithms (SIFT, SURF, ORB) [20] and Gracker algorithm [21] were selected to compare their time consumption, recognition loss rate, and average error with the research algorithm Ours.

Fig. 5. Comparison results of tilt, occlusion, small-scale, and illumination effects (first row: ORB, second row: Ours)

The results are shown in Table 1. When detecting in this scene, ORB and Ours recognition are faster, SURF and Ours recognition accuracy is higher, so the Ours algorithm is more suitable for use in this detection scene.

Because this study mainly focuses on reducing interference in actual situations during rapid detection, the ORB algorithm with the lowest time consumption was selected for comparison with the Ours algorithm. Seven sets of biomarkers were selected for comparison to intuitively compare the actual detection effect.

Fig. 6. Detection of Oblique Occlusion and Scale Change (ORB Detection Failure Omitted)

In general, there is not much difference in accuracy between the two algorithms, so this experiment mainly selects issues such as lighting, scale changes, and occlusion motion blur for comparison. As shown in Fig. 5 and Fig. 6, it can be seen that when facing situations with small tilted and occluded parts, both algorithms can also track successfully, but the accuracy of Ours is slightly higher than that of ORB; When the obstruction area of the landmark exceeds 50%, ORB tracking fails, while Ours tracking is good and has high accuracy; When faced with landmarks that only account for about 20% of the screen, ORB tracking fails, while Ours can track them with a small loss of accuracy. Considering the limited computing power of the processor, the accuracy of USB cameras is low, and the landmarks are relatively blurry when the scale changes greatly, it can be seen that Ours has a higher efficiency in processing landmarks under low computing power conditions. The effect of Ours real-time tracking five occluded landmarks under different lighting conditions is shown in Fig. 7. It can be seen that the Ours algorithm has high accuracy in real-time tracking of occluded environments with different lighting conditions.

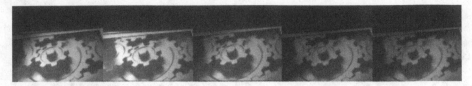

Fig. 7. Occlusion Detection Affected by Illumination

5   Conclusion

This article proposes a design method for optical flow tracking and positioning based on marker detection in intelligent assisted assembly scenarios. In order to overcome as much as possible the model jitter, occlusion, scale changes, and interference caused by metal material reflection caused by the camera during worker movement, a new algorithm framework is designed and constructed. Among them, the ORB algorithm is used to ensure ultra fast detection speed, and then the problem of occlusion and scale transformation is targeted by improving marker detection and pose solving. Finally, the pyramid optical flow algorithm is used to apply a marker tracking effect that can stabilize the model and is less affected by reflection to the overall detection framework. The experimental results show that the new framework can ensure high detection accuracy at a faster detection speed, and has good mobility effects for workers in intelligent assisted assembly scenarios, meeting the lightweight industrial needs under low computational power conditions.

References

1. Nistér, D., Naroditsky, O., Bergen, J.: Visual odometry. In: Proceedings of the 2004 IEEE Computer Society Conference on Computer Vision and Pattern Recognition 2004. CVPR 2004. IEEE, **1**, I-I (2004)
2. Davison, A.J., Reid, I.D., Molton, N.D., et al.: MonoSLAM: real-time single camera SLAM. IEEE Trans. Pattern Anal. Mach. Intell. **29**(6), 1052–1067 (2007)
3. Kitt, B.M., Rehder, J., Chambers, A.D., et al.: Monocular visual odometry using a planar road model to solve scale ambiguity. two thousand and eleven
4. Chien, H.J., Chuang, C.C., Chen, C.Y., et al.: When to use what feature? SIFT,SURF,ORB,or A-KAZE features for monocular visual odometry. In: 2016 International Conference on Image and Vision Computing New Zealand (IVCNZ). IEEE, pp. 1–6 (2016)
5. Mur-Artal, R., Montiel, J.M.M., Tardos, J.D.: ORB-SLAM: a versatile and accurate monocular SLAM system. IEEE Trans. Rob. **31**(5), 1147–1163 (2015)
6. Mur-Artal, R., Tardós, J.D.: Orb-slam2: an open-source slam system for monocular, stereo, and rgb-d cameras. IEEE Trans. Rob. **33**(5), 1255–1262 (2017)
7. Naixin, Q., Xiaogang, Y., Xiaofeng, L., et al.: A visual mileage calculation method based on ORB fFeatures and LK optical flow. J. Instrum. **39**(12), 216–227 (2018)
8. Jianjun, T., Bo, Y., Junhao, G.: Exploration and Practice of AR Intelligent guidance technology for aircraft assembly operations. Aviation Manufact. Technol. **62**(8), 22–27 (2019)
9. WESTERFIELD G,MITROVIC A,BILLINGHURST M.: Intelligent Augmented Reality Training for Motherboard Assembly. Int. J. Artif. Intell. Educ. **25**, 57–172 (2015)
10. Ran, L.: Research on pose estimation andstate detection methods for matrix components in AR assisted assembly [D]. Shanghai Jiao Tong University, Shanghai (2018)

11. Sun, J., Song, F., Ji, L.:https://ieeexplore.ieee.org/document/10046501/ " VINS-Mask: A
ROI-mask Feature Tracker for Monocular Visual-inertial System[J]. HYPERLINK " ROI-
mask Feature Tracker for Monocular Visual-inertial System[J].2022 International Conference
on Automation, Robotics and Computer Engineering (ICARCE).IEEE,2022:1–3
12. Harris, C.G., Stephens, M.J.: A combined corner and edge detector. Alvey vision conference
(1988)
13. Jiu, Y., Yangping, W., Jianwu, P., et al.: Improving the AR system long-term tracking
registration method for TLD and ORB. Comput. Eng. Appl. 57(7), 181 (2021)
14. Cvi š i ć I, Ć esi ć J,Markovi ć I,et al.: SOFT-SLAM: computationally efficient stereo visual
simultaneous localization and mapping for autonomous unmanned aerial vehicles. J. Field
Robot. 35(4), 578–595 (2018)
15. Persson, M., Piccini, T., Felsberg, M., et al.: Robust stereo visual odometry from monocular
techniques. In: 2015 IEEE Intelligent Vehicles Symposium (IV). IEEE, 2015 686–691 (2015)
16. Zhang, L., Curless, B., Seitz, S.M.: Rapid shape acquisition using color structured light
and multi-pass dynamic programming. In: International Symposium on 3D Data Processing
Visualization and Transmission Proceedings, pp. 24–36 (2002)
17. Hsieh C K,Lai S H,Chen Y C,2D expression-invariant face recognition with constrained
optical flow [C]‖2009 IEEE International Conference on Multimedia and Expo,June 28-July
3 2009,New York,NY,USA,New York: IEEE Press,2009;1058–1061
18. Ping, H., Zhen, C., Huan, W.: Binocular visual odometer based on circular feature matching.
J. Opt. 41(15), 3–4 (2021)
19. Wang, J.: Research on monocular vision odometer based on improved feature method [D].
China University of Mining and Technology (2021)
20. Yuren, H., Tiejun, L., Dong, Y.: Overview of three bit tracking and registration technology in
augmented reality. Comput. Eng. Appl. 55(21), 28–29 (2019)
21. Tao, W., Ling, H.: Gracker: a graph-based planar object tracker. IEEE Trans Pattern Anal.
Mach. Intell. 40(6), 1494–1501 (2017)

A Review of Image and Point Cloud Fusion
in Autonomous Driving

xiao ya Wang[✉], peng peng Zhang, meng shen Dou, and shu hao Tian

Shanghai Dianji University, Shanghai, China
1043274432@qq.com

Abstract. In the task of autonomous driving perception scenarios, multi-sensor fusion is gradually becoming the current mainstream trend. At this stage, researchers use multimodal fusion to leverage information and ultimately improve target detection efficiency. Most of the current research focus on the fusion of camera and LIDAR. In this paper, we summarize the multimodal-based approaches for autonomous driving perception tasks in deep learning within the last five years. And we provide a detailed analysis of several papers on target detection tasks using LiDAR and cameras. Unlike the traditional way of classifying fusion models, this paper classifies them into three types of structures: data fusion, feature fusion, and result fusion by the different stages of feature fusion in the model. Finally, it is proposed that the future should clarify the evaluation index, improve the data enhancement methods in different modes, and use multiple fusion methods in parallel in the future.

Keywords: Camera-LiDAR fusion · Object detection · Automatic driving · Multi-modal

1 Introduction

In order to navigate the vehicle safely and autonomously, it must be able to sense and interpret its surroundings in real time. This involves the use of a combination of sensors [1]. Advanced algorithms are used to process the extracted information, enabling the vehicle to react appropriately to changing situations, such as avoiding obstacles and adjusting speed. Environmental sensing and information extraction are therefore the key component of autonomous driving technology. The advancement of various technologies has made perception sensors such as camera and LiDAR increasingly common [2]. Perception tasks are significantly influenced by the raw data obtained from the environment through these sensors.

However, single-modal data usually has inherent drawbacks, such as occlusion and missing depth information [3]. LIDAR suffers from low resolution [4], sparseness, poor texture, and extreme weather conditions. Therefore, multimodal fusion using[1]LIDAR

---

[1] *Supported by National Natural Science Foundation of China Youth Fund (No.61802247) Natural Science Foundation of Shanghai (No.22ZR1425300) and Other projects of Shanghai Science and Technology Commission (No.21010501000).

J. Li et al. (Eds.): 6GN 2023, LNICST 554, pp. 62–73, 2024.
https://doi.org/10.1007/978-3-031-53404-1_6

and cameras can complement feature information for each other and enable better perception performance of self-driving vehicles.

In recent years, there has been significant development in multimodal fusion methods for autonomous driving target detection tasks [1]. Addressing the limitations of individual sensors, multimodal fusion combines data from multiple sensors to improve the accuracy and robustness of target detection. In this paper, we provide clear definitions and classification of different categories of multimodal fusion methods by analyzing relevant papers from the perspective of fusion stages over the last five years.

The main contribution of this work is to provide a comprehensive review of the latest research in multimodal fusion methods for autonomous driving target detection tasks. This review will help researchers to understand the current achievements and challenges in this field and inspire new ideas for future research. The main contribution of this work can be summarized as the following:

•To provide an advanced taxonomy for multimodal fusion methods in autonomous driving target detection tasks, we have identified and specifically defined three types of fusion: data fusion, feature fusion and result fusion. Each fusion method is explicitly defined based on the feature representation of the LiDAR and camera branches.
•We provide a detailed analysis of how various multimodal sensor fusion methods perform on the KITTI dataset.
•We analyze the current problems of multimodal fusion and introduce several potential research directions of these methods as well as future development directions.

2   Fusion Method

Based on multimodal fusion methods, the most commonly used approach is to integrate information from point clouds and images to achieve 3D detection. In this paper, we categorize these methods into three categories based on the stage of deep learning models where the feature fusion is performed: data fusion, feature fusion, and result fusion. Figure 1 illustrates the relationships among these categories.

2.1   Data Fusion

Data fusion is a technique which directly integrates individual modal data at the data level. Specifically, this approach refers to fusing image features with original point cloud data or fusing image features with point cloud features. The fused data is then utilized as input in subsequent models, without producing any candidate frames or regions of interest. In recent years, researchers have primarily adopted two fusion methods to combine RGB image and point cloud data. These methods involve attaching image features to original point cloud data or fusing image features with point cloud features.

The first type of fusion involves combining image features with point clouds. MVXNET [5], proposed by Vishwanath A. Sindagi in 2019, is an example of a fusion method where image features are directly attached to the point cloud. In this approach, the corresponding image of the point cloud is obtained and its features are extracted using a convolutional network before being attached to the point cloud. Other literature [6] combines features from the original point cloud with the image and then divides points

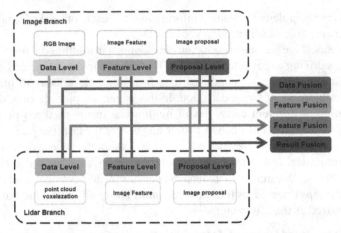

Fig. 1. Fusion Methodology Overview. We categorize multimodal fusion into three types: Data Fusion, Feature Fusion, and Result Fusion. Different combinations of LiDAR and camera data represent different fusion approaches. Feature fusion consists of two forms, which we distinguish with two colors.

in the scene into foreground and background points using that feature. This generates an initial 3D candidate frame for each foreground point.

In most research approaches, image features are fused in the form of image masks. For example, IPOD [7] performs semantic segmentation on 2D images and projects the segmentation result onto 3D point clouds, distinguishing front and background points. This allows point clouds to only detect front points when predicting proposals. Similar methods are also employed by PointPainting [8] and PI-RCNN [9]. PointAugmenting [10] modifies the point cloud by using point-like CNN features extracted from a pre-trained 2D detection model, and then performs 3D target detection on the reduced-dimensional point cloud. The literature [11] combines the image semantic segmentation results with LIDAR point cloud semantic segmentation results in the form of a BEV grid map. In 2023, PiMAE [12] developed a multimodal mapping module that aligns masked and visible tokens from two different modalities. This design highlights the critical role of the masking strategy in aligning tokens across multiple modalities. An example of fusion schematic using image masks can be seen in Fig. 2.

The second type of fusion is to fuse the point clouds and images after extracting features separately. The fusion schematic is shown in Fig. 3 In the cited works [13–16], the feature extraction process involves separately extracting features from point clouds and images, which are then fused together as the final output. The literature [17] adopts a fusion method that involves extracting features from point clouds in the form of bird's eye view (BEV) and projecting image features into the BEV space. The extracted image features and point cloud features at different resolutions are then fused using continuous convolution. The MSMDFusion [18] method promotes an intensive exploration of multi-level, multi-scale progressive interactions between LiDAR and camera features, utilizing a multi-scale architecture. 3D-CVF [19] uses auto-calibrated projection to convert the image features into a smooth spatial feature map with the highest correspondence to the

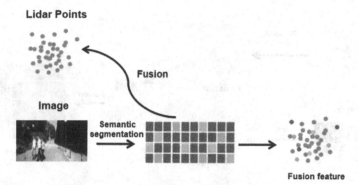

Fig. 2. An Example of Data Fusion I.This method assigns the semantic segmentation information from the image to the point cloud, allowing effective discrimination between foreground and background points. This enables the point cloud to only detect foreground points during proposal prediction.

point cloud features in the bird's eye view. The spatial attention mechanism is then used to fuse the features appropriately according to the region.

However, some of the height information is easily lost during the conversion of point clouds to bird's-eye view images. To eliminate this problem, the height information of the LiDAR point cloud is projected onto the RGB image and embedded to generate a new image called RGBD. EPNET [20] uses multiple LI-Fusion modules to fuse semantic image features point by point at different scales to improve the point features. In a similar manner, the study described in [21] utilizes a comparable approach by implementing PI-Fusion for point-by-point fusion.

SPLATNet [22] combines image and point cloud processing to improve target detection by supplementing point cloud information with image features. It also helps in semantic segmentation of the image by projecting point cloud information onto the image. Google introduces dynamic connectivity learning to enhance and attach geometric, semantic, and pixel texture information to 3D seed points selected from the point cloud. This process generates a pool of 3D object centers and the final 3D frame [23].

2.2 Feature Fusion

Feature fusion is an intermediate stage in the model where image and LIDAR-extracted features are fused. Unlike data fusion, feature fusion encompasses more than just stitching, spatial alignment, or projection of extracted features through their respective networks. Instead, one branch generates proposals and provides auxiliary information to predict the final result.

This type involves generating proposals for 3D targets on point cloud data and mapping them to multiple views. RoI features are cropped separately from the backbone networks of both the point cloud and image. Afterwards, the image RoI features are then fused with the point cloud RoI features.

In MV3D [24], 3D point clouds are projected to the top view and bird's-eye view. The latter is used to generate a low-precision 3D proposal after regression through a

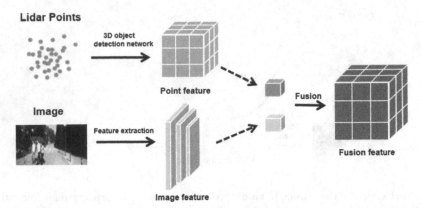

Fig. 3. An Example of Data Fusion II. The extracted features from the image sensor are integrated into the LiDAR data in order to enhance the LiDAR data.

convolutional network and a 3D bounding box. The proposal is then projected to the top view, bird's-eye view, and monocular image. And ultimately these features are fused and detected. A similar approach has been employed in literature [25]. AVOD [26] is an improved version of MV3D that simplifies the input aspect. It only utilizes point cloud aerial views and images with guaranteed results, and discards all the radar front views used in MV3D. This increases the speed of the process. MVX-NET [1] uses voxel processing for point clouds, dividing them into voxels. PI-RCNN [27] generates a proposal from the original point cloud through a 3D detection network. The semantic features are then fused directly with the image on the point cloud itself using the "PACF module." This module enhances the expressiveness of the fusion process. The schematic diagram of the first type of fusion model is shown in Fig. 4.

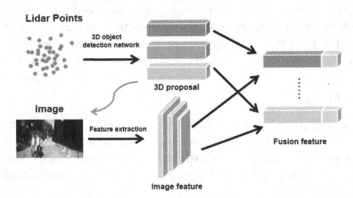

Fig. 4. An Example of Feature fusion I. 3D object proposals are generated based on the point cloud data and mapped onto multiple views. Combine the features of the image with the proposal of the point cloud. In the fusion feature, the point cloud's proposal plays a major role.

ImVoteNet [9] carries out separate processing of images and point clouds to obtain 2D frames and seed points. The fused information from the images and seed points generates 3D candidate frames by voting. In [28], "PAG-Net" is employed to generate precision and category-adaptive anchor points with a 2D target detector. A 2D target detector is employed to obtain category labels and 2D bounding boxes of objects. The resulting data is then projected onto 3D view cone space. Precision anchor points are generated from this projection to drive 3D object detection. In the referenced study [29], 2D objects are initially detected within images and cone-shaped ROI regions are generated based on geometric projection relationships. The point clouds within the ROI are then subjected to clustering. Figure 5 depicts a schematic of the second-class feature fusion.

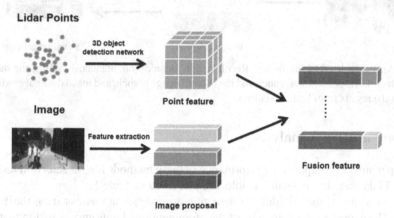

Fig. 5. An Example of Feature fusion II. The fusion features consist of features extracted from the point cloud and proposals generated in the image. The image proposals play a major role.

2.3 Result Fusion

Result fusion is the cross-modal fusion using detection results. The target detection task is accomplished by separately applying the base network to the image and point cloud data. Then, the detection results from the base network are used for decision-making to provide the initial position for optimizing subsequent bounding boxes. This approach eliminates the need to consider fusion and complementarity at the information level [30, 31]. The schematic diagram of result fusion is shown in Fig. 6.

In 2020, CLOCs proposed a technique [32] to extract images and point clouds after two branches for region suggestion. The proposed approach involved fusing the region suggestion candidates before non-maximum suppression (NMS) to obtain the final candidate frame. This helped in eliminating most of the redundant background regions to obtain more accurate detection results, which is known as result fusion for the candidate frame. A similar approach was adopted in [33–35].

Furthermore, a series of truncated cones are generated after RGB images [36–38]. The obtained truncated cones are used to group local points to obtain detection results. These methods heavily depend on the truncated cones generated by the images. Therefore, these methods are classified as result fusion as well. In addition, The study [39] uses result fusion in combination with other fusion methods to enhance recognition accuracy.

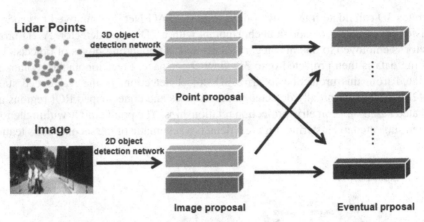

Fig. 6. An Example of Result fusion. Result-fusion methods simultaneously utilize the outputs of both the LiDAR point cloud branch and the camera image branch, and make the final prediction based on the results from both modalities.

3 Comparative Aanalysis

This paper aims to compare the performance of these methods for 3D detection tasks on the KITTI dataset. The detection results are presented in Table 1.

The majority of multimodal 3D detection approaches are assessed on the KITTI dataset. The experimental outcomes of the aforementioned multimodal fusion methods on the KITTI test dataset [40] are presented in Table 1. The performance of each method is evaluated across three categories, namely cars, pedestrians and bicycles, encompassing three levels of detection precision: easy, moderate, and difficult.

The data fusion model incorporates the semantic segmentation results of the image into the point cloud. This method achieves better detection performance compared to methods that directly fuse the point cloud image features. The semantic segmentation results complement the information of small target objects in the point cloud, thereby improving the detection results for pedestrians and bicycles. Among these methods, PointPainting yields the best performance.

The feature fusion model, AVOD, improves upon the effect of MV3D, achieving the best detection results among these methods. However, it is observed that the feature fusion model is less effective on the KITTI dataset.

The proposed result fusion model demonstrates minimal interaction between its branches, setting it apart from previous fusion techniques. The design of this model helps alleviate information interference and asymmetry issues that may arise during the interaction between different branches. The design of this model helps mitigate the information interference and asymmetry issues that can arise during interaction between different branches. By limiting the amount of interaction between branches, the model is less prone to being hampered by the poor detection performance of any one branch. Ultimately, this results in a more effective overall fusion detection process. However, due to a lack of mutual information supplementation, detection performance for small targets such as pedestrians and bicycles is relatively poor.

Table 1. Detection results of multimodal fusion models on the test set of KITTI dataset

Models	Car			Pedestrian			Cyclist		
	Easy	Mod	Hard	Easy	Mod	Hard	Easy	Mod	Hard
Data Fusion									
MVX-Net [5]	83.2	72.7	65.2	-	-	-	-	-	-
Poingnet&Refine [6]	**92.56**	83.16	80.52	-	-	-	-	-	-
IPOD [7]	79.75	72.57	66.33	56.92	44.68	42.39	71.4	53.46	48.34
PointPainting [8]	92.45	**88.11**	**83.36**	58.7	49.93	46.29	83.91	71.54	62.97
PI-RCNN [9]	84.37	74.82	70.03	-	-	-	-	-	-
PointFusion [12]	77.92	63	53.27	33.36	28.04	23.38	49.34	29.42	26.98
Complex-Retina [13]	78.62	72.77	67.21	-	-	-	-	-	-
Improved-RetinaNet [14]	-	-	-	55.16	47.82	41.91	-	-	-
Deep Parametric Continuous [16]	82.54	66.22	64.04	-	-	-	-	-	-
3D-CVF [19]	89.2	80.05	73.11	-	-	-	-	-	-
EPNET [21]	89.81	79.28	74.59	**94.22**	**88.47**	**83.69**	**96.13**	**94.22**	**89.68**
PI-Fusion [22]	91.73	81.49	76.13	-	-	-	-	-	-
Feature Fusion									
MV3D [25]	74.97	63.63	54.00	-	-	-	-	-	-
Cross-Modality [26]	**87.22**	**77.28**	**72.04**	-	-	-	-	-	-
AVOD [27]	81.94	71.88	66.38	50.80	42.81	40.88	64.00	52.18	46.61
Result Fusion									
CLOCs [32]	89.16	82.28	77.23	-	-	-	-	-	-
Real-time multisensory [34]	**96.02**	**89.05**	**88.38**	-	-	-	-	-	-
AVOD-MPF [36]	69.42	75.13	85.27	40.07	43.23	51.21	46.04	52.98	65.27
Frustum ConvNet [38]	89.02	78.8	77.09	-	-	-	-	-	-
Li module [39]	81.20	70.39	62.19	51.21	44.89	40.23	71.96	56.77	50.39

Data fusion detects better than the other two methods. This is because data fusion combines data from multiple modalities. Thus it improves the accuracy of detection. However, it has requirements on the degree of data matching. So it lacks flexibility. Compared to data fusion, Feature fusion combines features from various modalities, providing a flexible approach. Result fusion utilizes recognition outcomes of each modality. It compensates for the shortcomings of models that use predicted results from multiple branches. So it improve the effectiveness of modals. Overall, there is no one fusion method that works for all scenarios. The effectiveness of fusion methods is related on several factors, including the modalities fused, the data situation, and the network structure design. Therefore, it is necessary to choose the suitable fusion method according to the application scenario.

4 Trends and Promising Directions

Over recent years, there has been significant advancement in multimodal fusion methodologies for perception tasks related to autonomous driving. However, with the progress in this field, several new issues have arisen that require resolution.

•Calibrating multiple sensors and aligning data for fusion of point clouds with images may lead to information loss due to data variability.
•Data enhancement such as rotation and inversion to prevent overfitting cannot be performed at multiple sensors.
•Datasets and evaluation metrics are essential concerns that need to be addressed in multi-sensor fusion methods due to the limited availability of high-quality open-source datasets and evaluation metrics.
•Employing multiple fusion techniques is a future trend, albeit one that comes with increased complexity in model architecture and training.

Achieving accurate alignment in multi-sensor fusion methods often comes at a high cost. To address this issue, it is necessary to utilize surrounding information as a supplement to obtain better performance.

Additionally, projecting data during the dimensionality reduction of feature extraction can result in significant information loss. To mitigate this loss, mapping the two modal data to another high-dimensional representation designed for fusion can effectively utilize the original data in future work. Fusion can also be achieved by assigning weights to the values obtained from the multiple sensors. Future research should focus on mechanisms such as bilinear mapping to fuse features of different characteristics.

However, the problem of datasets and evaluation metrics is still a major challenge in multi-sensor fusion methods. To achieve better detection of the models, it is crucial to unify the evaluation metrics. This will also enable effective comparison between multiple models. Therefore, it is important to develop high-quality open-source datasets and evaluation metrics for future research in this field. Combining the three fusion methods introduces greater complexity to the model's architecture and training process. Further exploration is required to investigate how to simplify the model.

5 Conclusion

Multimodal fusion has emerged as a crucial trend in the perception task of autonomous driving. This paper aims to organize and classify various multimodal fusion methods employed in target detection for autonomous driving. These methods are categorized into three primary types of structures based on their fusion stages. Based on the analysis and comparison of the fusion stages, we have found that multimodal fusion techniques are highly effective in improving the accuracy and robustness of target detection in autonomous driving scenarios. The in-depth analysis and comparison of detection results on the widely-used KITTI dataset presented in this paper further highlight the efficacy of multimodal fusion methods.

But there are several challenges encountered in multimodal fusion, including information loss during fusion, the quality and availability of datasets, and limitations of evaluation metrics. Therefore, it is the trend to develop more advanced fusion techniques that can handle complex interactions among modalities. Moreover, using many kinds of fusion methods in models t is also one of the future research directions.

References

1. Yurtsever, E., Lambert, J., Carballo, A., Takeda, K.: A Survey of autonomous driving: common practices and emerging technologies. IEEE Access **8**, 58443–58469 (2020)
2. Wang, et al., "Multi-modal 3D Object detection in autonomous driving: a survey and taxonomy. IEEE Trans. Intell. Veh. (2023)
3. The DARPA urban challenge: autonomous vehicles in city traffic. springer (2009)
4. Xie, L., Xu, G., Cai, D., He, X.: X-view: non-egocentric multi-view 3D object detector. IEEE Trans. Image Process. **32**, 1488–1497 (2023)
5. Sindagi, V.A., Zhou, Y., Tuzel, O.: Mvx-net: Multimodal voxelnet for 3dd object detection. In: 2019 International Conference on Robotics and Automation (ICRA). IEEE pp. 7276–7282, (2019)
6. Zhen-dong, C.U.I., Zong-min, L.I., Shu-lin, Y.A.N.G., Yu-jie, L.I.U., Hua, L.I.: 3D object detection based on semantic segmentation quidance. J Graph., **43**(06), 1134–1142 (2022)
7. Yang, Z., Sun, Y., Shu, L., et al.: IPOD: Intensive Point-based Object Detector for Point Cloud.
8. 2019.Vora, S., Lang, A.H., Helou, B., et al.: Pointpainting: sequential fusion for 3d object detection. In: Proceedings of the IEEE/CVF Conference on Computer Vision and Pattern Recognition, pp. 4604–4612 (2020)
9. Xie, L., Xiang, C., Yu, Z., et al.: PI-RCNN: an efficient multi-sensor 3D object detector with point-based attentive cont-conv fusion module. In: Proceedings of the AAAI conference on artificial intelligence, 34(07), 12460–12467 (2020)
10. Wang, C., Ma, C., Zhu, M., et al.: Pointaugmenting: cross-modal augmentation for 3d object detection. In: Proceedings of the IEEE/CVF Conference on Computer Vision and Pattern Recognition. 2021: 11794–11803
11. Park, J., Yoo, H., Wang, Y.: Drivable dirt road region identification using image and point cloud semantic segmentation fusion. IEEE Trans. Intell. Transp. Syst. **23**(8), 13203–13216 (2021)
12. Chen, A., Zhang, K., Zhang, R., et al.: Pimae: point cloud and image interactive masked autoencoders for 3D object detection; In: Proceedings of the IEEE/CVF Conference on Computer Vision and Pattern Recognition, pp. 5291–5301 (2023)
13. Xu, D., Anguelov, D., Jain, A.: Pointfusion: deep sensor fusion for 3D bounding box estimation. In: Proceedings of the IEEE Conference on Computer Vision and Pattern Recognition, pp. 244–253 (2018)
14. Li, M., Hu, Y., Zhao, N., et al.: One-stage multi-sensor data fusion convolutional neural network for 3d object detection. Sensors **19**(6), 1434 (2019)
15. Li, W., Yuan, Q., Chen, L., Zheng, L., Tang, X.: Human target detection method based on fusion of radar and lmage data. Data Acquisit. Process. **36**(02), 324–333 (2021)
16. Tao, B., Yan, F., Yin, Z., Wu, D.:3D object detection based on high-precision map enhancement. J. Jilin Univ. (Engineering Edition) **53**(03), 802–809 (2023)

17. Jiao, Y., Jie, Z., Chen, S., et al.: MSMDfusion: Fusing lidar and camera at multiple scales with multi-depth seeds for 3d object detection. In: Proceedings of the IEEE/CVF Conference on Computer Vision and Pattern Recognition, pp. 21643–21652 (2023)
18. Liang, M., Yang, B., Wang, S., et al.: Deep continuous fusion for multi-sensor 3D object detection. In: Proceedings of the European Conference on Computer Vision (ECCV), pp. 641–656 (2018)
19. Yoo, J.H., Kim, Y., Kim, J., et al.: 3d-cvf: Generating joint camera and lidar features using cross-view spatial feature fusion for 3d object detection. In: Computer Vision–ECCV 2020: 16th European Conference, Glasgow, UK, August 23–28, 2020, Proceedings, Part XXVII 16. Springer International Publishing, pp. 720–736 (2020)
20. Wen, L.H., Jo, K.H.: Three-attention mechanisms for one-stage 3-d object detection based on LiDAR and camera. IEEE Trans. Industr. Inf. **17**(10), 6655–6663 (2021)
21. Huang, T., Liu, Z., Chen, X., et al.: Epnet: enhancing point features with image semantics for 3D object detection. In: Computer Vision–ECCV 2020: 16th European Conference, Glasgow, UK, August 23–28, 2020, Proceedings, Part XV 16. Springer International Publishing, pp. 35–52 (2020)
22. Cal, Z., Zhao, J., Zhu, F.: Single-stage object detection with fusion of point cloud and Image feature. Comput. Eng. Appl. **59**(09), 140–149 (2023)
23. Su, H., Jampani, V., Sun, D., et al.: Splatnet: sparse lattice networks for point cloud processing. In: Proceedings of the IEEE Conference on Computer Vision and Pattern Recognition, pp. 2530–2539 (2018)
24. Piergiovanni, A.J., Casser, V., Ryoo, M.S., et al.: 4d-net for learned multi-modal alignment. In: Proceedings of the IEEE/CVF International Conference on Computer Vision, pp. 15435–15445 (2021)
25. Chen, X., Ma, H., Wan, J., et al.: Multi-view 3d object detection network for autonomous driving. In: Proceedings of the IEEE conference on Computer Vision and Pattern Recognition, pp. 1907–1915 (2017)
26. Zhu, M., Ma, C., Ji, P., et al.: Cross-modality 3D object detection. In: Proceedings of the IEEE/CVF Winter Conference on Applications of Computer Vision, pp. 3772–3781 (2021)
27. Ku, J., Mozifian, M., Lee, J., et al.: Joint 3D proposal generation and object detection from view aggregation. In: 2018 IEEE/RSJ International Conference on Intelligent Robots and Systems (IROS). IEEE, pp.1–8 (2018)
28. Qi, C.R., Chen, X., Litany, O., et al.: Imvotenet: boosting 3D object detection in point clouds with image votes. In: Proceedings of the IEEE/CVF Conference on Computer Vision and Pattern Recognition, pp. 4404–4413 (2020)
29. Wu, Y., Jiang, X., Fang, Z., et al.: Multi-modal 3d object detection by 2D-guided precision anchor proposal and multi-layer fusion. Appl. Soft Comput. **108**, 107405 (2021)
30. Huang, Y., Li, B., Huang, Q., Zhou, J., Wang, L., Zhu, J.: Camera-LiDAR Fusion for Object Detection, Tracking and Prediction. J. Wuhan Univ. (Information Science Edition), pp. 1–8 (2023)
31. Cui, S., Jiang, H.-L., Rong, H., Wang, W.-Y.: A Survey of Multi-sensor Information Fusion Technology. Autom. Electron. **09**, 41–43 (2018)
32. Pang, S., Morris, D., Radha, H.: CLOCs: Camera-LiDAR object candidates fusion for 3D object detection. In: 2020 IEEE/RSJ International Conference on Intelligent Robots and Systems (IROS). IEEE, pp.10386–10393 (2020)
33. Wu, T.E., Tsai, C.C., Guo J.I.: LiDAR/camera sensor fusion technology for pedestrian detection. In: 2017 Asia-Pacific Signal and Information Processing Association Annual Summit and Conference (APSIPA ASC). IEEE, pp. 1675–1678 (2017)
34. Asvadi, A., Garrote, L., Premebida, C., et al.: Multimodal vehicle detection: fusing 3D-LIDAR and color camera data. Pattern Recogn. Lett. **115**, 20–29 (2018)

35. Zheng, S., Li, W., Hu, J.: Vehicle detection in the traffic environment based on the fusion of laserpoint cloud and image information. J. Instrument. Measure. **40**(12), 143–151 (2019)
36. Zhao, Y., Wang, X., Gao, L., Liu, Y., Dai, Y.: 3D target detection method combined with multi-view mutual projectiolfusion. J. Beijing Inst. Technol. **42**(12), 1273–1282 (2022)
37. Wang, Z., Jia, K.: Frustum convnet: Sliding frustums to aggregate local point-wise features for amodal 3D object detection. In: 2019 IEEE/RSJ International Conference on Intelligent Robots and Systems (IROS). IEEE, pp.1742–1749 (2019)
38. Qi, C.R., Liu, W., Wu, C., et al.: Frustum pointnets for 3D object detection from rgb-d data. In: Proceedings of the IEEE Conference on Computer Vision and Pattern Recognition, pp. 918–927 (2018)
39. Li, M.: Pedestrian Detection and Tracking Technology Based on the Fusion of Laser Point and Image [D].National University of Defense Technology (2017)
40. Guo, Y., Wang, H., Hu, Q., et al.: Deep learning for 3d point clouds: A survey. IEEE Trans. Pattern Anal. Mach. Intell. **43**(12), 4338–4364 (2020)

Deep Learning in Strawberry Growth Monitoring Research: A Review

Shuhao Tian$^{(\boxtimes)}$, Pengpeng Zhang, and Xiaoya Wang

Shanghai Dianji Univerisity, Shanghai, China
tsh_science@163.com
https://english.sdju.edu.cn/

Abstract. Intelligent equipment is increasingly employed in strawberry production to enhance fruit yield. To effectively monitor strawberry growth, the utilization of deep learning, specifically convolutional neural networks, has demonstrated remarkable effectiveness. This research paper delves into the study of deep learning techniques for monitoring strawberry growth and explores their applications in disease detection, fruit ripeness assessment, and fruit target identification. In addition, it provides an insightful analysis of the challenges encountered from both application and model perspectives. Furthermore, this paper proposes future trends, including the amalgamation of disease and fruit target detection, as well as the fusion of multiple algorithms.

Keywords: Deep Learning · Strawberry · Growth state

1 Introduction

China has consistently held the top position in strawberry cultivation area and production since 2007. Traditionally, strawberry growth monitoring has relied heavily on the expertise of agricultural professionals. However, this approach is susceptible to subjective biases, plant growth cycles, and environmental factors. With the advent of deep learning, experts have harnessed its potential for strawberry growth cycle monitoring. Leveraging deep learning algorithms, growth state monitoring has transformed into an efficient and automated process, encompassing crop planting, fruit picking, and evaluation. By leveraging data such as images and nutritional indicators, deep learning methods enable precise monitoring and evaluation of strawberry growth. Deep learning has gained significant traction in various fields, including crop classification, yield assessment, fruit development tracking, and pest and weed detection, showcasing impressive outcomes.

Supported by National Natural Science Foundation of China Youth Fund(No. 61802247), Natural Science Foundation of Shanghai(No.22ZR1425300), and Other projects of Shanghai Science and technology Commission(No.21010501000).

J. Li et al. (Eds.): 6GN 2023, LNICST 554, pp. 74–80, 2024.
https://doi.org/10.1007/978-3-031-53404-1_7

This paper presents research on deep learning applications in strawberry growth monitoring. It provides a comprehensive overview of the current state of deep learning technology applied to strawberry ripeness detection, pest and weed identification, plant nutrient status tracking and prediction, and fruit harvest location and marking. By analysing the benefits and challenges of deep learning, this review identifies potential research directions for its application in strawberry growth monitoring, providing valuable insights to drive future intelligent advances in strawberry cultivation.

2   Research Progress

Deep learning, an advanced approach based on multilayer neural networks, revolutionizes the extraction of data features. In the context of strawberry growth detection, deep learning proves instrumental in tackling challenges related to crop pest and weed detection, fruit recognition, and soil quality analysis. By harnessing the power of deep learning, agricultural automation production can achieve higher levels of efficiency, thereby fostering the growth and development of strawberry cultivation practices. Deep learning in growth monitoring workflow includes, image acquisition, image preprocessing, model design, model training, and recognition analysis. Figure 1 shows the workflow of deep learning in strawberry monitoring.

This section serves as an introduction to key aspects of strawberry growth monitoring, encompassing disease detection, weed damage assessment, fruit ripeness evaluation, and fruit target detection. Additionally, the characteristics of traditional machine learning and deep learning approaches are analyzed in detail.

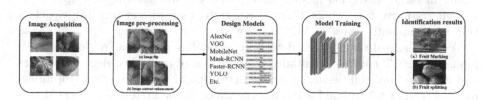

Fig. 1. Strawberry monitoring workflows

2.1   Disease

The detection and management of plant pests and diseases are crucial aspects of agricultural production. Deep learning techniques, particularly convolutional neural networks (CNNs), have emerged as effective tools for classifying disease-infected crop images from healthy ones.

Noteworthy studies include Cheng Dong [1] proposed an innovative operator integrated into the fully connected layer of ALexNet, resulting in improved disease recognition compared to the original model. Aguirre Santiago et al. [2] employed EfficientNet and SSD target detection algorithms, achieving a disease detection accuracy of 81% for three strawberry diseases. Irfan Abbas [3] proposed multiple models, including ALexNet and EfficientNet-B3, to address the low efficiency of manual detection of strawberry leaf scorch disease. Experimental results demonstrated that EfficientNet-B3 achieved the highest classification accuracy. Jaemyung Shin [4] and Muhab Hariri hariri2022tipburnutilized ResNet and PSO-CNN models, achieving recognition accuracies exceeding 98% for disease image recognition.

2.2   Strawberry Ripeness

Assessing the ripeness of strawberries is a crucial factor in determining optimal harvest conditions. Deep learning has proven to be a valuable tool for accurate strawberry ripeness assessment.

Xin Li et al. [6] utilized the Otus algorithm in combination with CaffeNet to evaluate the ripeness of strawberry fruit images, achieving a remarkable recognition rate of 95%.In a similar vein, Mauricio Rodriguez et al. [7] developed a lightweight convolutional neural network model based on the VGG-16 architecture, achieving a ripeness classification accuracy of 96.34%. Building upon this progress, Wan Hyun Ch [8] employed a combination of VGG-16 and MLP models, leading to outstanding fruit ripeness assessment results with a classification accuracy of 90%.

2.3   Strawberry Target Detection

Automated agricultural picking systems have gained significant attention in the field of agricultural automation. The study of strawberry ripeness and advancements in target detection models have paved the way for the development of strawberry harvesting robots. Current research in strawberry fruit target detection predominantly relies on YOLO.

YOLO, a real-time target detection algorithm proposed by Joseph Redmon [9] in 2015, enables simultaneous prediction of multiple object boundaries and category probabilities using neural networks. M G Dorrer et al. [10] implemented YOLO V3 to build a strawberry robot vision system for target detection and fruit labeling, achieving a recognition accuracy of 93%. To address the challenge of capturing clear nighttime images, Youchen Fan [11] and Liu Xiaogang [12] enhanced nighttime fruit recognition using YOLO V5, incorporating dark channel enhancement and gamma transform methods, resulting in a recognition accuracy of 90%. Other studies, such as those by Jun Sun [13] and Madis Lemsalu [14], improved average accuracy values by adding GhostNet and utilizing YOLO V5 for detecting fruit and pedicel positions, achieving recognition accuracies of 92.62% and 91.5%, respectively. Meanwhile, experts proposed Faster R-CNN [15] and Unet [16] to identify fruits and achieve good results according to the application scenarios.

Table 1. Deep learning model applications and performance metrics

Year	Model	Research	Improvement (Accuracy %)
2021	AlexNet + improved operator [1]	Disease	3.78
2021	EfficientDet-D0 [2]	Disease	5
2021	EfficientNet-B3 [3]	Disease	11
2021	SqueezeNet-MOD1 [4]	Disease	4.3
2022	PSO-CNN [5]	Disease	15
2018	CaffeNet [6]	Maturity	11
2021	VGG-16 + max pooling [8]	Maturity	23
2021	VGG-16+MLP [7]	Maturity	8
2020	YOLO V3 + Relu [10]	Object detection	3
2020	YOLO V3 + Gamma correction [12]	Object detection	20.15
2022	YOLO V5 + Dark Channel [11]	Object detection	20
2022	YOLO V4-Tiny [13]	Object detection	5.77
2022	YOLO V5s [14]	Object detection	10
2018	Faster R-CNN [15]	Object detection	35
2022	U-Net + same [16]	Object detection	2

2.4 Analysis

Deep learning surpasses traditional machine learning by addressing the limitations in capturing image features. By employing convolutional and pooling layers, deep learning excels in capturing a broader range of features. In the context of the intricate growth environment of strawberries, deep learning proves to be more effective. Table 1 illustrates the research directions and models of deep learning used in monitoring the growth of strawberries. The primary applications of strawberry growth monitoring encompass the detection of strawberry diseases, weed damage, fruit ripeness, and fruit targets.

The current trend in disease modeling focuses on lightweight approaches to enable prolonged crop monitoring, allowing for better disease observation. To gather information effectively, farms extensively employ drones and small IoT devices. However, deep learning models typically involve a large number of parameters and demand high-performance resources. In contrast, lightweight models have fewer parameters and lower equipment requirements. Consequently, current research primarily emphasizes the development of lightweight models.

In fruit target detection, the application of the model is directed towards image enhancement and improving the real-time.

(1) The strawberry robot can work all day. However, the low light at night will reduce the robot's acquisition efficiency. Therefore, image quality can be improved by means of image enhancement to ensure model recognition accuracy.

(2) Fast models enhance robot acquisition efficiency. So, experts improve the model backbone network and feature extraction unit to increase the model running speed.

Numerous factors contribute to the challenges of accurate and timely strawberry monitoring. These reasons can be summarized as follows: similar growing conditions, fruit shading, camera sensitivity to light, and image noise. Specifically:

(1) The detection and identification of fruits and diseases require a rapid response capability. And model complexity is influenced by the natural environment, such as light exposure, strawberry growth cycles and weed proliferation. However, these models usually need more time and have poor real-time.
(2) Plant diseases can undergo changes in appearance and color as they develop, while the intricate natural environment contributes to varying degrees of disease progression. Moreover, multiple diseases may exhibit similar symptoms, making early intervention by agricultural experts challenging.

The future direction of strawberry monitoring will focus on two key areas: lightweight models and complex background improvement. The lightweight approach aims to reduce model parameters without sacrificing recognition performance, thereby enhancing operational speed. Many researchers have already made progress in developing lightweight models to achieve this goal. Complex background improvement can be achieved through image pretreatment and unstructured transformations. Image pre-processing techniques can reduce image noise and address challenges such as fruit overlap. Unstructured modifications involve implementing measures like adding shade cloths to the acquisition environment and optimizing foliage, among other methods, to enhance the quality of environment acquisition.

3   Conclusion

The importance of strawberry growth monitoring has increased with the expanding market demand for strawberries and the advancements in automated agriculture. Deep learning has gained popularity among scholars due to its high recognition accuracy, making it a prominent direction in agricultural growth monitoring. This paper provides a comprehensive overview of research on deep learning technology in strawberry growth monitoring. It categorizes and analyzes research directions and models, highlighting the differences between traditional machine learning and deep learning in this domain, ultimately concluding that deep learning offers superior performance.

The objective of strawberry growth monitoring is to enhance planting efficiency and mitigate losses resulting from factors such as diseases. Currently, disease detection and fruit picking are treated as separate directions. However, integrating the detection of diseases and fruit picking will prove more effective in strawberry management through the use of strawberry robots. Therefore, achieving model generalization, incorporating multiple features, and leveraging various algorithms will be crucial technological breakthroughs in the field of strawberry monitoring. Further research and development of strawberry growth monitoring technology are imperative.

References

1. Dong, C., Zhang, Z., Yue, J., Zhou, L.: Classification of strawberry diseases and pests by improved alexnet deep learning networks. In: 2021 13th International Conference on Advanced Computational Intelligence (ICACI), pp. 359–364. IEEE (2021)
2. Santiago, A., Solaque, L., Velasco, A.: Strawberry disease detection in precision agriculture. In: Proceedings of the 18th International Conference on Informatics in Control, Automation and Robotics - ICINCO, pp. 537–544. INSTICC, SciTePress (2021)
3. Abbas, I., Liu, J., Amin, M., Tariq, A., Tunio, M.H.: Strawberry fungal leaf scorch disease identification in real-time strawberry field using deep learning architectures. Plants 10(12), 2643 (2021)
4. Shin, J., Chang, Y.K., Heung, B., Nguyen-Quang, T., Price, G.W., Al-Mallahi, A.: A deep learning approach for RGB image-based powdery mildew disease detection on strawberry leaves. Comput. Electron. Agric. 183, 106042 (2021)
5. Hariri, M., Avşar, E.: Tipburn disorder detection in strawberry leaves using convolutional neural networks and particle swarm optimization. Multimedia Tools Appl. 81(8), 11795–11822 (2022)
6. Li, X., Li, J., Tang, J.: A deep learning method for recognizing elevated mature strawberries. In: 2018 33rd Youth Academic Annual Conference of Chinese Association of Automation (YAC), pp. 1072–1077. IEEE (2018)
7. Rodriguez, M., Pastor, F., Ugarte, W.: Classification of fruit ripeness grades using a convolutional neural network and data augmentation. In: 2021 28th Conference of Open Innovations Association (FRUCT), pp. 374–380. IEEE (2021)
8. Cho, W.H., Kim, S.K., Na, M.H., Na, I.S.: Fruit ripeness prediction based on DNN feature induction from sparse dataset. CMC-Comput. Mater. Contin 69, 4003–4024 (2021)
9. Redmon, J., Divvala, S., Girshick, R., Farhadi, A.: You only look once: Unified, real-time object detection. In: Proceedings of the IEEE Conference On Computer Vision and Pattern Recognition, pp. 779–788 (2016)
10. Dorrer, M., Popov, A., Tolmacheva, A.: Building an artificial vision system of an agricultural robot based on the darknet system. In: IOP Conference Series: Earth and Environmental Science, vol. 548, p. 032032. IOP Publishing (2020)
11. Fan, Y., Zhang, S., Feng, K., Qian, K., Wang, Y., Qin, S.: Strawberry maturity recognition algorithm combining dark channel enhancement and yolov5. Sensors 22(2), 419 (2022)
12. Xiaogang, L., Cheng, F., Jianian, L., Yanli, G., Yuyang, Z., Qiliang, Y.: Strawberry recognition method based on convolutional neural network. J. Agricult. Mach. 51(2), 237–244 (2020)
13. Jun, S., Yide, C., Xin, Z., feng, S.J., hong, W.X.: An improved yolo v4 tiny model for fast and accurate identification of strawberries in sheds. J. Agricult. Engi. 38(18), 9 (2022)

14. Lemsalu, M., Bloch, V., Backman, J., Pastell, M.: Real-time CNN-based computer vision system for open-field strawberry harvesting robot. IFAC-PapersOnLine **55**(32), 24–29 (2022)
15. Lin, P., Chen, Y.: Detection of strawberry flowers in outdoor field by deep neural network. In: 2018 IEEE 3rd International Conference on Image, Vision and Computing (ICIVC), pp. 482–486. IEEE (2018)
16. Jia Z W, Yao S M, Zhang R Y, et al.: Image segmentation of strawberry in greenhouse based on improved u-net network. J. Shanxi Agricult. Univ. (Natural Science Edition) **42**(2), 120–128 (2022)

Image, Video, and Signal Processing & Software Engineering

Stepwise Change and Refine Network for Human Pose Transfer

Han Mo🆔, Yang Xu(✉)🆔, Youju Peng🆔, and Guidong Xu🆔

College of Big Data and Information Engineering, Guizhou University,
Guiyang 550025, China
xuy@gzu.edu.cn

Abstract. Achieving person image synthesis through pose guidance is an arduous and profound endeavor. Addressing the limitations of previous approaches, namely inaccurate pose generation and inconsistencies with the target texture, we present a new network. This two-stage network is designed not only to reposition a given person's image to the intended pose but also to produce outcomes that are more believable and closely mirror authentic images. In the initial stage of our network, we employ Coarse Blocks, a series of modules sharing a uniform structure, to generate coarse images. This process involves gradually transforming the raw images towards the target pose, fostering improved shape consistency. Subsequently, we extract style features from the reference image using semantic distribution. These features play a pivotal role in optimizing the coarse image, leading to the production of the final generated image. Significantly, this enhanced image demonstrates increased fidelity to the visual characteristics of the target image. To further enhance the perceptual quality of the generated images, we introduce a novel loss function. This loss function is instrumental in aligning the generated images more closely with our cognition. We compare and ablate experiments with current state-of-the-art models, attest to the remarkable improvements achieved by our model, as evidenced by significant enhancements in SSIM, PSNR, and LPIPS scores.

Keywords: Deep learning · image generation · human pose transfer · person image synthesis

1 Introduction

Synthesizing person images is a practical and formidable undertaking within the realm of image generation. We need to use human image synthesis in many application scenarios, The network proposed in this paper is the application of the human pose transfer in person image synthesis. Example is shown in Fig. 1.

When performing pose transfer, a crucial step involves the analysis and extraction of keypoint information from the original pose, which is subsequently

Supported by Guizhou Provincial Key Technology R&D Program [2023] General 326.

J. Li et al. (Eds.): 6GN 2023, LNICST 554, pp. 83–95, 2024.
https://doi.org/10.1007/978-3-031-53404-1_8

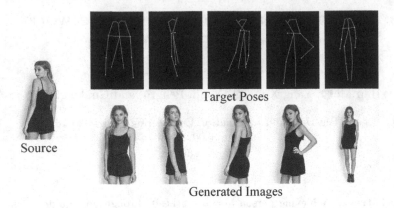

Target Poses

Source

Generated Images

Fig. 1. A example of human pose transfer.

projected onto the desired target pose. Despite extensive research and application of pose transfer technology in computer vision, there remain numerous obstacles and issues to address. One such challenge arises from potential instances of missing or mismatched keypoints during the pose migration process. To address this, Huang et al. [1] proposed a method called PATN, which utilizes an attention mechanism to gradually guide the pose and restrict the range of pose changes in each transfer step. Although PATN has shown promising results in this area, the gradual transfer process may result in the progressive loss of the source image's style features. Consequently, subsequent works such as those by Men et al. [2], and Zhang et al. [3] have introduced modules specifically designed to extract reference style features. These works fuse the style features extracted from the source image with the parsing map to generate the final image, effectively preserving the clothing textures. However, the ability to map the spatial relationship of the pose may be compromised when different body parts are occluded.

To deal with aforementioned problems, our proposed model seeks to uphold the coherence between the generated image and the original image, while systematically transitioning the pose. Our method facilitates improved shape uniformity between the generated and authentic images. Initially, the pose keypoints and conditional semantic distribution are fed into a sequence of modified PATB modules (based on Zhu et al.'s progressive attention tree branch) to generate an initial output with coarse details. Subsequently, we employ pose features and textures features to refine and optimize the initial coarse output, leading to the final result. It is crucial to optimize both the initial and final generated results. This process is essential for optimizing and refining the generated images to better align with the objectives. This ensures the continuous improvement and alignment of the generated images with the desired objectives.

We have identified several key contributions to our paper, and here's a brief summary:

- We present a novel model, a effective approach to pose transfer that combines two key ideas: incrementally producing transfer outcomes and transferring stylistic attributes utilizing a parsing map. Our method is designed to generate the spatial relationship of the posture more accurately while retaining the style features of the original image. By gradually generating transfer results, we ensure that the final output is realistic and closely aligned with human perception.
- We introduce a variant of the PATB model called Coarse Block. We aim to better constrain the changes in posture during the pose transfer process. Additionally, we enhance the learning ability of the network by introducing the attention mechanism after each block. This module enables the model to capture and incorporate both global and local information, further improving the quality of the generated images.
- In our work, we introduce the use of an $L1$ loss specifically on the face part during the pose transfer process. By focusing on the face, we aim to produce more realistic and visually appealing results.

2 Related Works

Human pose transfer becomes more and more popular in recent years. Huang et al. [4] proposed a pose transfer method based on style transfer, which uses Adaptive Instance Normalization (AdaIN) technique to transfer the style from the reference image to the generated image. The Deformable Convolutional neural Network (DCN) module can learn the intricate adjustments required for specific regions during generation. By incorporating the groundbreaking "deformable G-convolution" technique, this innovative approach effectively handles entities with intricate forms, such as human body parts, producing individuals that are lifelike and natural. Huang et al.'s attention-based pose transfer method [4] incorporates multiple submodules with identical structures. Through progressive training, this method produces more realistic pose transfer images, successfully addressing intricate pose transfer tasks. While the portrait images generated by Huang et al. rely on pose keypoints, resulting in occasional discontinuities between keypoints in output images, particularly when significant pose changes occur, XingGAN introduces an alternative discriminator architecture called the "average pooling discriminator." This advancement greatly enhances the model's capacity to comprehend both overarching and nuanced details.

3 Method

We aim to enhance the consistency between the original image and the target pose by utilizing our proposed model. The model is divided into two parts, as demonstrated in Fig. 2. In the first stage, we generate images, denoted as I_{crs}, that align with the original pose P_r. The second stage aims to refine the initially generated image I_{crs} to achieve a closer resemblance to the target image I_t.

To accomplish this, we utilize as inputs the reference image I_r, the reference parsing map S_r, the reference pose P_r, and the target pose P_t.

The dataset contains a series of images depicting the same person in diverse poses, along with detailed pose information. Specifically, the pose representation consists of 18 human keypoints extracted using the Human Pose Estimator, and the parsing map S_r is derived from the Part Grouping Network [5], as provided by [1,3].

3.1   Coarse Block

Inspired by PATB [1], we propose the Coarse Block (CB). Initially, we combine S_r, P_r, P_t, and S_g obtained via prediction. Subsequently, after convolution processing, this leads to the feature F_p^0 that functions as the first input for CB. We also integrate Criss-Cross Attention into the module which allows for better spatial information capture without introducing a significant computational burden.

Consider the t-th block, The inputs to the CB module consist of F_p^{t-1}, representing the pose code, and F_i^{t-1}, representing the image code. By passing through the convolutional layer, I_r is transformed into the feature F_i^0. This process is illustrated in Fig. 3. The input of each block is the output of the previous block, and divided into two pathways, called pose pathway and image pathway respectively. The convolutional modules of two pathways use the same construction with

The attention matrix M is obtained by employing the Criss-Cross Attention mechanism. The procedure can be delineated as follows:

$$M = \sigma \left(CCA \left(Conv_p \left(F_p^{t-1} \right) \right) \right) \tag{1}$$

where CCA stands for Criss-Cross Attention function. The update equations for F_i^{t-1} and F_p^{t-1} can be expressed as follows:

$$F_i^t = M \odot Conv_{crs} \left(F_i^{t-1} \right) + F_i^{t-1} \tag{2}$$

In the equations given, the symbol $\odot$ denotes element-wise multiplication. To obtain F_p^t, we concatenate F_p^{t-1} and F_i^t.

3.2   Refine Block

The main problem addressed by the Refine Block is how to parse I_{crs} using S_r, this step aims to refine and align the generated image with the expected structure in order to enhance consistency with the ground truth image.

3.3   Training

The optimization of the network involves incorporating various components into the loss as follows:

$$\mathcal{L} = \lambda_{l1}\mathcal{L}_{l1} + \lambda_{adv}\mathcal{L}_{adv} + \lambda_{per}\mathcal{L}_{per} + \lambda_{par}\mathcal{L}_{par} + \lambda_{face}\mathcal{L}_{face} \tag{3}$$

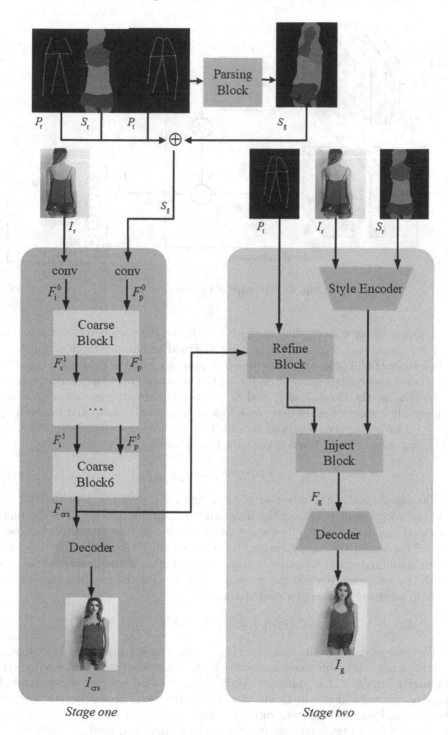

Fig. 2. Overview of our network.

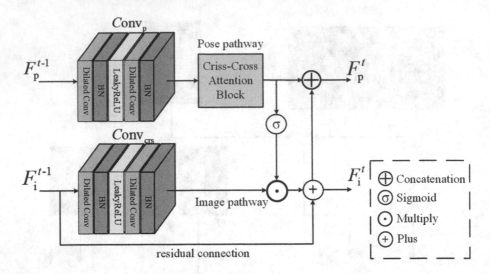

Fig. 3. Structure of Coarse Block.

To ensure a faster convergence of the network, weight parameters (denoted as λ_{cor}, λ_{l1}, λ_{adv}, λ_{per}, λ_{par}, and λ_{face}) are assigned to balance the loss values of each component. These weight parameters play a crucial role in adjusting the importance and contribution of each part to the overall optimization process.

To increase the likeness of I_g and I_t, pixel-level similarity is measured using L1 losses. It is worth noting that these losses are not only computed between I_g and I_t, but also between I_{crs} and I_t. This helps to improve the generation of I_g in the first stage. The $L1$ loss is described as follows:

$$\mathcal{L}_{l1} = \|I_g - I_t\|_1 + \|I_{crs} - I_t\|_1 \tag{4}$$

The calculation of the adversarial loss relies heavily on the discriminator D. This loss component penalizes the divergence in distribution between the final generated image I_g and the first stage generated image I_{crs}, as well as between both of them and the target image. By incorporating the adversarial loss, we aim to minimize the differences in distribution between the generated images and the target, ensuring that the generated output closely resembles the desired image in terms of its overall visual characteristics.

$$\mathcal{L}_{adv} = \mathbb{E}\left[\log\left(1 - D\left(I_g\right)\right)\right] + \mathbb{E}\left[\log\left(1 - D\left(I_{crs}\right)\right)\right] + \mathbb{E}\left[\log D\left(I_t\right)\right] \tag{5}$$

The perceptual loss function, drawing on the concept pioneered by Johnson et al. [6], gauges the differences in feature representations between the generated and target images. This approach emphasizes the similarity of the generated result to the ture image in terms of visual perception, preserving the overall structure and semantic content of the original image. Rather than solely focusing on pixel-level matching, the perceptual loss takes into account higher-level features. The formula for the perceptual loss can be expressed as:

$$\mathcal{L}_{per} = \|\Phi\left(I_g\right) - \Phi\left(I_t\right)\|_1 \tag{6}$$

where Φ denotes AlexNet.

The loss of parsing map is computed using $L1loss$, which can be represented by the following formula:

$$\mathcal{L}_{par} = \|S_g - S_t\|_1 \tag{7}$$

To address the difficulty in generating high-quality face images during training, we incorporate a specialized loss function specifically designed for faces. First, we utilize the FaceBoxes algorithm to generate and save the bounding box coordinates of the detected faces. Then, we compute the loss by cropping the corresponding regions from both the generated image I_g and the target image I_t using these bounding box coordinates. This allows us to focus on improving the quality and accuracy of the generated faces. The formula for this face-specific loss can be expressed as follows:

$$\mathcal{L}_{face} = \|C\left(I_g\right) - C\left(I_t\right)\|_1 \tag{8}$$

where C denotes the function used to crop the face.

4   Setup

4.1   Dataset

Our training process incorporates the In-shop Clothes Retrieval Benchmark, a subset of the extensive DeepFashion dataset proposed by Liu et al. (2016) [7]. This benchmark encompasses a vast collection of 52,712 high-resolution fashion images featuring models. The dataset provides paired images that showcase the same individual wearing identical clothing items. To facilitate our training process, we follow the dataset splits recommended by Zhu et al. (2019) [1], which assist in appropriately partitioning the data into training, validation, and evaluation sets. These measures enhance the accuracy and reliability of our model's performance. In the training set, there are a grand total of 101,966 pairs, and in the testing set, there are 8,570 pairs.

4.2   Metrics

The Similarity Index Measure (SSIM) is a frequently employed evaluation metric that gauges the structural likeness between two images by considering factors such as luminance, contrast, and structure, taking into account both local and global image characteristics. SSIM calculates the similarity index by considering the mean, variance, and covariance of the compared images. Higher SSIM values indicate a higher level of similarity between the images, while lower values suggest greater dissimilarity. The SSIM metric has proven to be effective in assessing image quality and evaluating image processing algorithms. Peak Signal to Noise Ratio (PSNR) value indicates difference between the generated image and the real image. A lower value for Fréchet Inception Distance (FID) means that there is less difference between the generated image and the real data distribution, so the generator performs better. Compared with other indicators, FID can better capture the overall features of an image, rather than just local details, so it is widely used in evaluating GAN models. The LPIPS loss leverages neural networks to grasp the feature representations necessary for gauging image similarity based on human subjective perception.

5   Experiment

5.1   Contrast Experiment

Quantitative Comparison. During the evaluation phase, we conduct a comprehensive comparison between our model and several state-of-the-art models, including PATN [1], ADGAN [2], PINet [8] and PISE [3]. To ensure fairness and consistency, the evaluation images used in this comparison are provided by the respective authors. These images have a fixed width and height of 256×176. Consequently, the dimensions of 256×256 may lead to cropping of the generated image in certain models.

In our comprehensive evaluation, to ensure a fair and unbiased comparison, we utilize the same dataset split, allowing us to measure the evaluation metrics on the entire test set. The quantitative results are presented in Table 1. The smaller the FID and LPIPS, the better, while the larger the PSNR and SSIM, the closer the image quality is to the ground and truth.

From the table, it is evident that our model exhibits a slight improvement in SSIM, indicating its ability to accurately predict details, texture, structure, and other image aspects compared to the other models. In addition, our model achieves the highest scores on the PSNR and LPIPS metrics, which further support its alignment with human visual perception. Although slightly lower than the best performer, our FID score still demonstrates the high quality and similarity between the generated and real images. The positive outcomes affirm the superior performance of our proposed method compared to existing models, showcasing its potential for various image generation tasks.

Table 1. Quantitative comparisons.

Network	FID	PSNR	SSIM	LPIPS
PATN	17.1699	15.5684	0.7069	0.2861
ADGAN	12.2656	16.7704	0.7459	0.2255
PINet	11.4476	16.9132	0.7415	0.2167
PISE	**11.0102**	16.9081	0.7417	0.2078
Ours	11.1308	**17.2853**	**0.7492**	**0.1888**

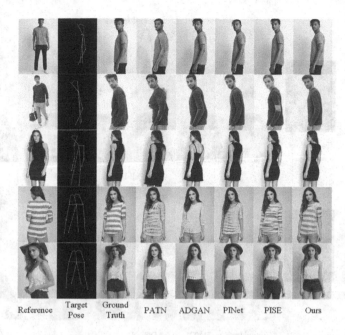

Reference   Target   Ground   PATN   ADGAN   PINet   PISE   Ours
             Pose   Truth

Fig. 4. Qualitative comparisons.

Qualitative Comparison. Figure 5 showcases the generated results of our model in comparison to state-of-the-art networks, namely PATN [1], ADGAN [2], PINet [8], and PISE [3]. It is important to note that all the displayed images have been released by their respective authors.

In the example depicted in the last row of Fig. 5, both PATN and ADGAN fail to capture the feature of hat, resulting in a generated portrait without the hat. This issue highlights a limitation of the attention mechanism employed in PATN, as it focuses only on a specific region of the image and may overlook crucial details in certain cases. Furthermore, some of the images generated by PISE exhibit inaccuracies in correctly generating the boundary between clothes and skin. These comparisons emphasize the advantages of our proposed method, as it effectively addresses the limitations associated with attention mechanisms and normalization modules. As a result, our model consistently produces more accurate and visually appealing generated images across various scenarios.

However, our model's results clearly demonstrate that it excels in generating images with precise structures while effectively preserving crucial texture features. It is important to note that when producing full-body images, the facial area is relatively small in comparison to the entire image. In other models, this often leads to blurry or distorted facial features. However, our model, thanks to the inclusion of L_{face}, manages to generate facial features that remain reasonable even in such scenarios. For a specific example, please refer to Fig. 6. Additionally, upon reviewing all the images generated using the test set, it is evident that the quality of our generated images remains consistently high and stable.

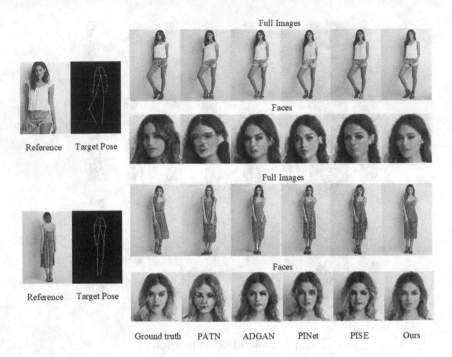

Fig. 5. Face comparisons

Ablation Study. We conducted experiments on several ablation networks to evaluate the effectiveness of our proposed component.

Less Coarse Blocks (3-CBs). In this section, we will investigate the impact of reducing the number of Coarse Blocks in our model. The Coarse Blocks are responsible for progressively changing a person's posture, and their structure is relatively simple, relying on attention mechanisms to drive the process. In our original model, we utilized six Coarse Blocks to achieve the desired progressive posture change. However, to explore the potential trade-off between performance and computational efficiency, we conducted experiments with a modified version of our model that only consists of three Coarse Blocks. Through this experiment, we observed that reducing the number of Coarse Blocks led to a noticeable decrease in the quality of the generated images. The generated results exhibited less accurate and realistic posture changes compared to the full model with six Coarse Blocks. This indicates that the additional Coarse Blocks contribute significantly to the overall performance of the model in capturing and transforming postures effectively.

While using three Coarse Blocks reduced the computational complexity of the model, it came at the cost of compromising the quality and fidelity of the generated images. Therefore, we recommend utilizing the original configuration with six Coarse Blocks for achieving the best results in terms of both posture progression and image quality.

Without Refine Block (w/o RB). The Refine Block plays a crucial role in our model by selectively retaining and incorporating style features into specific areas where they are needed. This block takes as input the target pose and the generated parsing map, which encode the desired appearance and structure of the final image. The Refine Block performs fusion manipulation between the encoded result and the initially crudely generated image. This process allows for the refinement and enhancement of the generated image based on the encoded information. By selectively injecting style features from the encoded result, the Refine Block helps to improve the overall quality and fidelity of the generated image, ensuring that it aligns more closely with the desired target pose and appearance.

Without L_{face} (w/o L_{face}). In the complete model, the face portion of both the true image and the generated image was precisely isolated using identical bounding box frames. This ensured that the L_{face} loss could be computed, thereby facilitating the generation of a more plausible facial representation. However, in this section, the utilization of the L_{face} loss is omitted by the model.

Full Model. All components proposed were utilized in the training of our model.

To ensure consistency, all our ablation models were trained using the same settings. Table 2 provides a quantitative comparison of the performance of each model, while Fig. 6 presents a qualitative comparison. Our analysis indicates that the model containing all the components obtained the best results across all evaluation metrics. It is interesting to note that the omission of certain loss functions resulted in a considerable decrease in the performance of the models. Furthermore, we observed a pronounced difference in the visual quality of the generated images between the models with and without certain loss functions, suggesting that these losses play an essential role in guiding the generation of high-quality images.

Table 2. Quantitative comparisons.

Network	FID	PSNR	SSIM	LPIPS
3-CBs	11.3343	17.1161	0.7341	0.1944
w/o RB	12.3178	16.3635	0.7286	0.2193
w/o L_{face}	11.2032	17.2145	0.7425	**0.1840**
Full Model	**11.1308**	**17.2853**	**0.7492**	0.1888

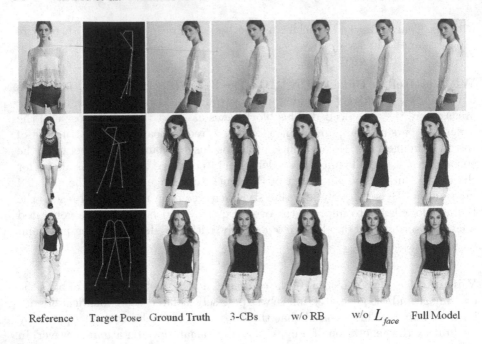

Reference   Target Pose Ground Truth   3-CBs   w/o RB   w/o L_{face} Full Model

Fig. 6. Qualitative comparisons in ablation study.

Our model consists of three CBs, but its performance in terms of LPIPS is below-par. This indicates that the model may lack efficient space transformation abilities due to the reduced number of CBs. It is important to note that using more CBs could potentially lead to better results. Additionally, we observed that the model without a Refine Block exhibited significantly lower performance across all metrics compared to the full model. Finally, by analyzing the qualitative comparison, we observed that the absence of a Refine Block leads to an increased dependence of the shape of the generated image I_g on I_{crs}. While the model without L_{face} exhibits minimal variance from the baseline in terms of evaluation metrics and even surpasses the baseline slightly in terms of LPIPS, the face produced by the full model outperforms that generated by the model without L_{face}. We acknowledge your perspective that the evaluation metrics may not show a substantial difference between models due to the relatively small proportion compared to the whole picture. It is plausible that the impact of the face region on calculating the evaluation metrics might be relatively less significant. However, it is important to consider that the evaluation metrics are designed to capture overall performance and provide a quantitative measure of the model's abilities. While the face area might be relatively small, it still plays a crucial role in generating realistic and visually accurate facial representations. Therefore, even if the impact on evaluation metrics might not be as pronounced, ensuring the fidelity of the face region remains essential for generating high-quality results.

6   Conclusion

This paper introduces a pioneering model for human pose transfer, which is designed to maintain the spatial structure and style characteristics. Our approach comprises two distinct parts, each fulfilling a specific purpose. In the first stage, we employ multiple modules with similar structures to gradually alter the pose and generate an initial sketch. This iterative process ensures the preservation of the spatial structure throughout the pose transformation. Moving on to the second stage, we further enhance the generated image by incorporating keypoints and semantic distribution information. Additionally, we introduce style features to refine former stage output. To improve the quality of the facial image, we include facial loss calculations as well. Through extensive experimentation, we demonstrate that our approach produces visually appealing results that closely resemble real images. In conclusion, our paper introduces an innovative model for human pose transfer that successfully maintains both pose structure and texture features. The experimental results support the visual fidelity of our approach, while the ablation study provides further validation of its effectiveness.

References

1. Zhu, Z., Huang, T., Shi, B., Yu, M., Wang, B., Bai, X.: Progressive pose attention transfer for person image generation. In: Proceedings of the IEEE/CVF Conference on Computer Vision and Pattern Recognition, 2019, pp. 2347–2356 (2019)
2. Men, Y., Mao, Y., Jiang, Y., Ma, W.Y., Lian, Z.: Controllable person image synthesis with attribute-decomposed gan. In: Proceedings of the IEEE/CVF Conference on Computer Vision and Pattern Recognition, 2020, pp. 5084–5093 (2020)
3. Zhang, J., Li, K., Lai, Y.K., Yang, J.: PISE: Person image synthesis and editing with decoupled gan. In: Proceedings of the IEEE/CVF Conference on Computer Vision and Pattern Recognition, 2021, pp. 7982–7990 (2021)
4. Huang, X., Belongie, S.: Arbitrary style transfer in real-time with adaptive instance normalization. In: Proceedings of the IEEE International Conference on Computer Vision, 2017, pp. 1501–1510 (2017)
5. Gong, K., Liang, X., Li, Y., Chen, Y., Yang, M., Lin, L.: Instance-level human parsing via part grouping network. In: Proceedings of the European Conference on Computer Vision (ECCV), 2018, pp. 770–785 (2018)
6. Johnson, J., Alahi, A., Fei-Fei, L.: Perceptual losses for real-time style transfer and super-resolution. In: Computer Vision-ECCV,: 14th European Conference, Amsterdam, The Netherlands, October 11–14, 2016, Proceedings, Part II 14. Springer **2016**, 694–711 (2016)
7. Liu, Z., Luo, P., Qiu, S., Wang, X.,Tang, X.: Deepfashion: Powering robust clothes recognition and retrieval with rich annotations. In: Proceedings of the IEEE Conference on Computer Vision and Pattern Recognition, 2016, pp. 1096–1104 (2016)
8. Zhang, J., Liu, X., Li, K.: Human pose transfer by adaptive hierarchical deformation. In: Computer Graphics Forum, vol. 39, no. 7. Wiley Online Library, 2020, pp. 325–337 (2020)

Performance Analysis of Web Server Side Reactive Programming

Haojie Li$^{(\boxtimes)}$ and Xu Guo

School of Electronics and Information, Shanghai DianJi University, Shanghai, China
lhj_work@126.com

Abstract. The traditional imperative programming paradigm at web server side encounters some performance limitations when handling lots of concurrent requests. The performance bottleneck cannot be resolved solely by greater hardware investment and program optimization, which will raise system operating expenses. As a new programming paradigm, reactive programming can handle many concurrent requests with a small number of threads, thereby boosting system performance. This paper proposes benchmarks for performance analysis of reactive web applications: TPS (Transaction Per Second) and ART (Average Response Time), CPU utilization, and network bandwidth utilization, and quantitatively analyzes the performance of imperative application and reactive application in various scenarios. According to the experimental results, the reactive programming paradigm can increase TPS by 6.5 times and decrease ART by about 93% when compared to the imperative programming paradigm in a Linux server cluster with 10 Intel i5–, 12-core CPUs, 4GB memory, 10MB/S network bandwidth, and a maximum of 10,000 requests. Network bandwidth utilization is decreased by around 2 times, while CPU resources are lowered by about 3 times. This demonstrates that reactive programming can enhance the performance and reliability of web applications while handling heavy concurrent demands more effectively.

Keywords: Reactive Programming · Performance Analysis · Web Applications

1 Introduction

With the rapid development of the Internet, both the demands and the usage frequencies of web applications are growing rapidly. The design architecture of web applications has evolved from a monolithic architecture to distributed architecture. Then SOA (Service Oriented Architecture) appears, until today, the prevalent micro-service architecture. The optimization of software architecture enhances the performance of web applications inherently [1].

SOA divides and reassembles the business logic of the distributed architecture. But there are still bottlenecks when massive concurrent requests arrive [2, 3]. The micro-service architecture has lately gained popularities, each service has a distinct function, its own data collection, and a unique life cycle, enabling fine-grained scaling to

J. Li et al. (Eds.): 6GN 2023, LNICST 554, pp. 96–108, 2024.
https://doi.org/10.1007/978-3-031-53404-1_9

adapt to changing workloads [4]. However, the micro-service architecture must maximize deployment and have a high reliance on resources. The micro-service architecture uses a synchronous invocation mechanism, which prevents each service from freeing up resources while waiting for a response. This can lead to extreme waste of system resources in high concurrency scenarios [5]. Additionally, the synchronous imperative programming paradigm serves as the foundation for the micro-service architecture. The main thread must block to wait for the data processing task or events processing to complete. To sum up, the micro-service architecture also faces bottlenecks in improving system performance that are difficult to break through.

Reactive programming is an asynchronous data flow-based programming paradigm. It can obtain response data not only in event-driven mode, but also in data-driven mode. During the process of acquiring response data, the main thread is not required to block, resulting in greater efficiency and stability [6]. As an asynchronous and non-blocking programming paradigm, reactive programming has the potential to become a valuable supplementary to the micro-service architecture. As far as the authors' knowledge, there is no literature on performance analysis of reactive web applications, particularly from a quantitative perspective.

This article analyzes the performance of reactive web applications. The authors propose the performance testing benchmarks of reactive programming: throughput, average response time, CPU utilization, and network bandwidth utilization. Experiments have demonstrated that reactive programming can significantly boost performance and decrease resource consumptions.

Related Works

This paper concentrates on the quantitative performance analysis of reactive web applications. It summarizes the relevant research works into two aspects: the study of reactive programming technology and the performance analysis of web applications.

Reactive programming. Bainomugisha et al. compared and analyzed reactive and imperative programming in six dimensions, concluded that reactive programming was more adept at handling data processing and concurrency performance issues [6]. Salvaneschi et al. examined 127 observer pattern program objects. In comparison to the traditional object-oriented programming paradigm, reactive programming can substantially enhance program comprehension [7]. Mogk proposed a fault tolerance method for reactive programming in distributed systems [8]. Without modifying program functions, this method can automatically store program states and recover program operation in the event of a catastrophe, while ensuring fine-grained state consistency. Literature [9] examined the efficacy of Java reactive programming. The aforementioned researches concentrate on the reactive programming paradigm, but not the server-side, which is the focus of this study.

Performance evaluation. Comprehensive performance analysis of web applications is the most effective method to improve website performance and resource utilization, and different web applications have different performance testing standards [10]. For applications that provide underlying services, Ghanavati et al. examined memory leakage problems and proposed a classification method for leakage types, defects that contribute to leakage, and corrective measures [11]. Riet et al. quantitatively evaluated the efficacy of target web applications using benchmark testing tools [12]. For performance analysis, the authors consider that the most important testing objects for web applications under high concurrent conditions should be throughput, response time, and network bandwidth consumption.

2 Background Techniques

2.1 Reactive Programming

Reactive programming is a programming paradigm that considers data streams as continuously changing sequences and provides rich APIs and data transformation operators. The central concept of reactive programming is reactive streams, which entails subscribing to and processing data streams using the observer pattern [7]. The method of data stream processing is asynchronous and non-blocking. Synchronization and asynchronization refer to the manner in which a process or thread transmits a request and awaits a response from another process or thread. In the synchronous mode, the thread that sends the request awaits the response result before continuing with subsequent operations, whereas in the asynchronous mode, it returns immediately after sending the request and receives the response result via a callback function [13].

When performing I/O operations, blocking and non-blocking refer to whether a process or thread waits for an operation to complete before continuing execution. In blocking mode, when a process or thread conducts I/O operations, if the operation is not completed, the process or thread will wait until the operation is complete before executing subsequent operations. In non-blocking mode, if an I/O operation is not completed when a process or thread performs I/O, the process or thread will promptly return and continue to execute subsequent operations. In subsequent operations, polling or event notification are used to ascertain whether the I/O operation has been completed [14].

Reactive Programming is comprised of three fundamental ideas: Observer Pattern, Stream, and Back Pressure.

Observer Pattern
Observer pattern refers to a dependency relationship between multiple objects with a one-to-many dependency ratio. When the state of an object changes, all dependent objects are promptly notified and updated. Occasionally, this pattern is also known as the publish-subscribe pattern or the model-view pattern [15]. The main components of the observer pattern are depicted by Fig. 1.

(1) Subject: Likewise known as the abstract target class. It provides an aggregate class for storing observer objects, addition and deletion methods for observer objects, and an abstract method for notifying all observers.
(2) Concrete Subject: It implements the abstract target's notification method. When a concrete subject's internal state changes, it notifies all registered observer objects.
(3) Observer: This interface defines an update mechanism for all concrete observers. This interface's purpose is to promptly update itself after receiving the subject's notification.
(4) Concrete Observer: Implements the update interface defined by the abstract observer role to coordinate its state with the state of the subject. The concrete observer role can, if necessary, store a reference to the concrete subject role.

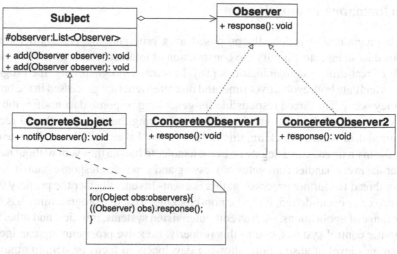

Fig. 1. Observer Pattern

Stream

Stream is an abstract concept constructed on the foundation of functional programming. Each stream flow's operation process follows the pattern of creating→ operating→ obtaining results, similar to how assembly line nodes form chains. Each stream operation will generate a new stream and utilize delayed fetching. Only when the user requires the result will its execution occur [16].

Back Pressure

In an asynchronous scenario, back pressure is a strategy that instructs the upstream observer to slow down the sending rate when the downstream observer transmits events faster than the upstream observer can process them. The working principle of back pressure is shown in Fig. 2.

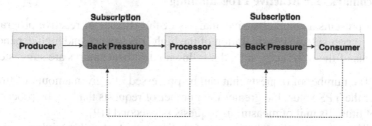

Fig. 2. Back Pressure

As Fig. 2 shows, when the producer transmits data too quickly, the downstream consumer notifies the Processor via the backpressure strategy, and the Processor notifies the upper-level producer via the backpressure strategy after receiving the request in order to slow down the data sending rate. Back pressure is not an operator that can be used directly in a program, it is merely a strategy for controlling the rate of event flow [17].

3 Performance Evaluation

Reactive programming is initially proposed as a programming paradigm for asynchronous data streams to simplify the construction of interactive user interfaces and the drawing of real-time system animations [18]. In reactive programming, the program's behaviors and data both evolve over time, and discrete events are processed in the order in which they occur. The thread responsible for generating response data notifies the main thread via an event when the data is available, ensuring that the main thread receives the required data without blocking, this is data-driven. Likewise, when a discrete event occurs, the main thread can delegate an event handler to handle the event without halting, and after the event handler completes processing and generates response data, it notifies the main thread for further processing, this is event-driven. Whether the primary thread is data-driven or event-driven, it can be non-blocking. Reactive programming has found a wider range of applications, such as computer vision systems, robotics, and other types of computer control systems, due to this property. Reactive programming can increase the program's level of abstraction, allowing developers to focus on defining the event dependency relationships of business logic without having to worry about the specifics of event handling, resulting in cleaner code.

However, it has not been studied whether separating event dependency relation and event handling details can improve program performance. In light of this issue, this paper analyses the efficiencies of the reactive web application (using WebFlux as an example) and the imperative web application (using WebMVC) in various application scenarios as an illustration. The experiment focuses on the performance of reactive and imperative servers in blocked and non-blocked states in response to queries with high concurrency.

In this paper, all testing is conducted in an isolated environment (virtual machine). The testing server is a cluster of 10 Linux machines with Intel i5–10400@2.90GHz 12-core processors, 4GB memory, and 8000kbps network bandwidth, running Ubuntu 22.10 and Oracle JDK 11. The testing tool is JMeter 5.5, which is used to simulate requests with high concurrency and capture data on performance indica-tors.

3.1 Benchmarks for Reactive Programming

This paper presents a comprehensive analysis of the efficacy of reactive programming on the web server side. In consideration of the high concurrency and high dependability requirements of the web server side, the following test benchmarks are suggested:

(1) TPS: The number of requests that can be processed in a given amount of time. The greater the TPS value, the greater the number of requests that can be processed per unit of time. Its unit of measure is requests per second [19].
(2) ART: The amount of time between submitting a request and receiving a re sponse. The shorter the response time for a request, the lower the ART value. Its unit is milliseconds [20].

(3) CPU Utilization: Refers to the CPU usage during request processing, which re flects the stability of the system [21].
(4) Network Bandwidth Utilization: Refers to the speed at which data can be transmitted from one network node to another within a specified time frame. Typically, it is measured by the quantity of data transmitted per second [22].

On the web server side, the keys to success are high throughput and quick response time. This paper evaluates the efficiencies of the two applications by comparing their load-dependent throughput and average response time. In order to simulate a scenario in which a significant number of web requests utilize cloud services, the network bandwidth utilization is introduced as a test benchmark. The experiment compares the network bandwidth utilization of the two applications over a period of time under different levels of traffic and congestion. Hardware resources, especially CPU utilization, are essential for the continuous operation of the web applications. Within 5 min, the CPU utilization under varying loads and fixed blocking as well as the number of requests are recorded in order to compare the hardware resources being occupied.

4 Experimental Results and Analysis

4.1 TPS and ART

TPS and ART are crucial performance indicators for web applications. In this experiment, the efficacies of WebFlux and WebMVC is assessed by comparing their TPS (Transactions Per Second) and ART (Average Response Time). To obtain the TPS and ART values, it is essential to continuously increase the load on the tested systems. The increasing system load is represented as a gradual increase in concurrent requests. Let NoR represents the number of loads, CT (Concurrent Threads) represents the number of concurrent threads, and LC (Loop Count) represents the number of cycles, there is $NoR = CT*LC$.

Notably, in order to mimic a high-blocking environment, this paper actively sets the program's delay time which makes the program delay a certain amount of time after sending each request to simulate the time it spends waiting for system resources. This brings the system to the precise level of expected load.

The test code fragments are listed as follows: where WebFlux is a reactive application, while WebMVC is a traditional imperative application.

```
//WebFlux
@RestController
  @RequestMapping("/webflux")
  public class AsyncController {
      @GetMapping("/latency/{param}")
      public Mono<String> hello(@PathVariable long param)
{
            return Mono.just("this is async web return")
                    .delayElement(Duration.ofMillis(param);
      }
  }


//WebMVC
@RestController
  @RequestMapping("/normal")
  public class NormalWebController {
      @GetMapping("/latency/{param}")
      public String delay(@PathVariable long param){
          try {
              TimeUnit.MILLISECONDS.sleep(param);
          } catch (InterruptedException e) {
              return  "Error during thread sleep";
          }
          return "this is normal web return";
      }
  }
```

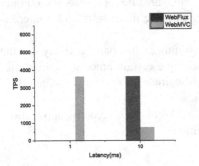 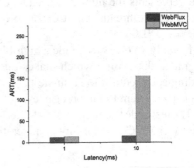

Fig. 3. TPS and ART

The first round of testing simulates the efficacies of WebFlux and WebMVC as the number of requests increases. The experiment sets the number of concurrent threads to 200, the number of LC to 100, and the load to 20,000. The blocking time is 1ms and 10ms, respectively, and Fig. 3 compares TPS and ART of the two applications.

The second round of testing simulates the efficacies of WebFlux and WebMVC under various traffic conditions in a high-blocking environment. There are a total of five sets of loads, with each set ranging from 2,000 to 10,000, and the blocking duration is set to 100 ms. Experimental results for TPS and ART are shown by Fig. 4.

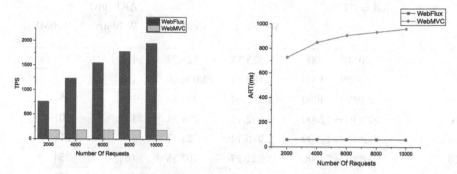

Fig. 4. Experimental results for TPS and ART

As the above experimental results show, the performance of WebFlux in the presence of high concurrency and high blocking is significantly superior to that of WebMVC.

In the aforementioned investigations, all requests are generated concurrently, and the asynchronous nature of reactive programming distinguishes it from conventional programming paradigms. Different applications based on the two paradigms invoke different numbers of threads for the same number of requests. This round of experiment regulates the number of threads and cycle times in order to generate different loads.

In addition, this test controls various loads and blocking times of the system in a unified environment, and observe the overall performance of the tested system.

In the first two phases, the effects of load and blocking time on performance are examined separately, but not simultaneously. This round of experiment considers the simultaneous variation of loads and blocking time, and investigate the effect on performance.

Table 1 provides more specific experimental results.

As the results shown in Table 1, it can be seen that WebFlux performs significantly better than WebMVC in terms of TPS and ART when blocking time and loads both rise. TPS and ART are both influenced by NoR(loads) and blocking time. Changes to the loads and blocking time have a comprehensive effect on TPS and ART, and do not indicate that one factor is significantly more influential than the other.

More specifically, the values of TPS and ART are jointly affected by loads and blokcing time. Increasing load can reduce TPS and enhance ART, and increasing blocking time can have the same effect. These two factors influence the system's performance indicators through the same mechanism, and there is no evidence that one factor overpowers the other completely.

When the blocking time is fixed to 1ms, and the load is 4000, the TPS growth rate of WebFlux increases by 6.42 percent compared to the load of 1000, but ART does not improve. When the load is 4000 and the blocking time is 10ms, the TPS growth rate of WebFlux increases by 54.3% and the ART increases about 6.72 times. When the blocking time is 40 ms, and the load is 40,000, the TPS growth rate of WebFlux increases

Table 1. Comparison of TPS and ART under different conditions

Conditions		Results			
Blocking time (ms)	NoR = CT*LC	TPS		ART (ms)	
		WebFlux	WebMVC	WebFlux	WebMVC
1	100*10 = 1000	525.17	523.75	1	1
1	200*20 = 4000	1,182.31	1,096.09	1	1
10	200*20 = 4000	1,207.04	745.78	11	85
10	400*30 = 12000	3,642.11	739.34	11	101
20	400*30 = 12000	2,916.74	721.13	15	211
20	600*40 = 24000	5,921.27	707.38	16	254
40	600*40 = 24000	4,754.43	693.81	29	582
40	800*50 = 40000	8,012.04	691.32	32	678
80	800*50 = 40000	6,780.10	702.26	42	728
80	1000*60 = 60000	8,907.33	704.17	45	824
100	1000*60 = 60000	9,523.77	709.12	62	1056
100	1500*70 = 105000	10,113.47	699.37	67	1293

about 4.72 times while the growth rate of ART increases about 1.12 times compared to the load of 24,000. A detailed comparison of the results can be seen in Table 1.

The results above can directly demonstrate that WebFlux performs far better than WebFlux under conditions of high blocking and high concurrency, and they can also demonstrate indirectly how important asynchronous and non-blocking properties of reactive programming can be for web applications. Compared to imperative web applications, reactive web applications are more powerful to handle the difficulties of high blocking and high concurrency settings.

4.2 CPU Utilization

CPU utilization is an essential indicator during system operations, particularly in the case of high concurrency. High CPU utilization by a single service not only weaken the computing power seriously, but also raises the hardware temperature and shortens the CPU's lifespan. Consequently, reducing the performance of the system. The experiment studies how the number of concurrent threads and their duration affect CPU utilization. The experiment determines that the number of concurrent threads is 200, the blocking time is 1ms, and Fig. 5 depicts the CPU utilization rate when the load is maintained for 5 min.

Figure 5 demonstrates that when the number of threads and blocking time are fixed, the CPU utilization of WebFlux is always less than that of WebMVC within 5 min, and the difference is about 10%. Subsequently, 6 sets of tests are set up with an increasing

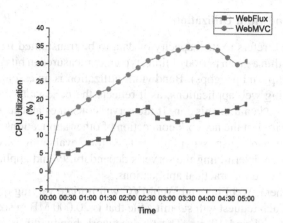

Fig. 5. Variations in CPU utilization within 5 min

number of concurrent threads, with 10 tests per group. The recorded results of the CPU utilization rates of WebFlux and WebMVC are listed in Table 2.

Table 2. Within five minutes, the CPU utilization of concurrent threads with a 1 ms blocking time rises.

NoR = CT*LC	CPU (%)		Reduced Proportion (%)
	WebFlux	WebMVC	
100*10 = 1000	5	4	25
150*20 = 3000	7	13	46.1
200*30 = 6000	12	23	47.8
250*40 = 12000	13	35	62.8
300*50 = 15000	15	40	62.5
350*60 = 21000	17	58	70.7

The CPU utilization of both WebFlux and WebMVC increases with the number of threads, as shown in Table 2. However, the CPU utilization of WebFlux is always below 20%, while the CPU utilization of WebMVC reaches a maximum of 58%, which is 3.4 times that of WebFlux.

Because WebMVC adopts a synchronization mechanism, as the number of concurrent requests increases, blocking will inevitably occur, which will result in more threads consume system resources. Reactive programming is an asynchronous mechanism, which can call fewer threads to complete the requests, so it does not occupy too much CPU, and can release system resources in time, thus achieve better performance.

4.3 Network Bandwidth Utilization

Network bandwidth refers to the capacity of data to be transmitted from one network node to another within a given period of time, typically measured in bits per second (bps) or its multiples (kbps, mbps, gbps). Bandwidth utilization is an essential performance metric for evaluating web applications, as it reflects the bandwidth resources utilized by the server during data transmission. If the server consumes an excessive amount of bandwidth, it may impact the network connection of other applications and even lead to network congestion and transmission failure. Therefore, evaluating network bandwidth occupancy can aid in determining the server's dependability and applicability in order to make the best choice for practical applications.

In this experiment, WebFlux and WebMVC will initiate varying quantities of concurrent requests. Each request will submit a file that is 20,000 MB in size. The blocking time are specified to 1 and 10 ms. The total network bandwidth is 8,000 kbps. The experimental outcomes are depicted in Fig. 6.

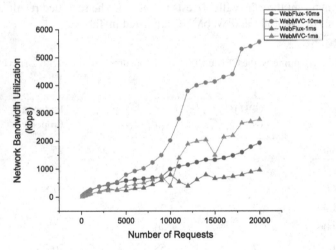

Fig. 6. Network Bandwidth Utilization under different number of concurrent requests

Figure 6 reveals that when the blocking time is 1 ms and the total number of requests (load) is less than 10,000, the network bandwidth occupancy curves of WebFlux and WebMVC are nearly identical. When the load increases steadily from 10,000, the network bandwidth utilization of WebMVC increases dramatically. The network bandwidth utilization of WebMVC at a load of 20,000 is approximately 2,500 kbps, or 31.25 percent of the total network bandwidth. WebFlux's highest network bandwidth utilization is 970 kbps, which accounts for 12.12% of the total network bandwidth. WebMVC's highest network bandwidth utilization is approximately 2.57 times that of WebFlux.

As shown by Fig. 6, when blocking time is 10ms, the number of loads of 4700 is a turning point. There is no discernible difference between the two subjects when the blocking time is 10ms and the load is below 4,700. As the number of loads increases, the network bandwidth utilization of WebMVC increased dramatically to 5,500 kbps, reaching 68.75% of the total network bandwidth. The bandwidth utilization of WebFlux is 2,500 kbps, which reaches 31.25% of the total network bandwidth. The network bandwidth utilization of the former is approximately 2.2 times that of the latter.

This demonstrates that under high concurrency and a large amount of data transfer, the network bandwidth occupancy of reactive applications is significantly lower than that of imperative applications, thereby occupying fewer network resources and decreasing the likelihood of network congestion and transmission failure.

5 Conclusion

This paper proposes testing benchmarks for reactive web applications, including throughput and average response time, CPU utilization, and network bandwidth utilization under varying request loads, and analyzed quantitatively the efficacies of WebFlux and WebMVC in various scenarios.

First, in terms of throughput and response time, when the blocking time is 100 ms and the number of loads is 60,000, the throughput of a reactive web application increases approximately 12 times and the response time decreases about 16 times. This demonstrates that reactive programming is more suitable for handling high concurrency requests.

In terms of CPU utilization, reactive programming is considerably more efficient than imperative web applications. When the load (total number of current requests) reaches 21000, the CPU utilization rate decreases 3.4 times. If the program continues to run for 5 min, the CPU utilization is 10% lower than that of imperative web applications. This demonstrates that reactive programming makes more efficient use of computing resources.

Finally, in terms of network bandwidth occupation, reactive programming uses only one-third of network bandwidth resource as much as that of imperative web applications.

In conclusion, the application of reactive programming technology on the web server-side can considerably improve the performance of web applications, particularly in the processing of a large number of concurrent requests and the utilization of network bandwidth in highly blocking environments. The research results provide empirical support for the application of reactive programming technology in the field of web applications and introduce new concepts and techniques for improving performance.

Exploring the application of reactive programming in other fields, such as mobile applications and the Internet of Things, may be the future work.

References

1. Kornienko, D.V., Mishina, S.V., Melnikov, M.O.: The single page application architecture when developing secure Web services. J. Phys. Conf. Ser. **2091**, 012065 (2021)
2. Hamad, Z.J., Zeebaree, S.R.M.: Recourses utilization in a distributed system: a review. Int. J. Sci. Bus. **5**, 42–53 (2021)
3. Niknejad, N., Ismail, W., Ghani, I., Nazari, B., Bahari, M., Hussin, A.R.B.C.: Understanding service-oriented architecture (SOA): a systematic literature review and directions for further investigation. Inf. Syst. **91**, 101491 (2020)

4. Li, S., et al.: Understanding and addressing quality attributes of microservices architecture: a Systematic literature review. Inf. Softw. Technol. **131**, 106449 (2021)
5. Hannousse, A., Yahiouche, S.: Securing microservices and microservice architectures: a systematic mapping study. Comput. Sci. Rev. **41**, 100415 (2021)
6. Bainomugisha, E., Carreton, A.L., van Cutsem, T., Mostinckx, S., de Meuter, W.: A survey on reactive programming. ACM Comput. Surv. **45**, 1–34 (2013)
7. Salvaneschi, G., Proksch, S., Amann, S., Nadi, S., Mezini, M.: On the positive effect of reactive programming on software comprehension: an empirical study. IIEEE Trans. Software Eng. **43**, 1125–1143 (2017)
8. Mogk, R., Baumgärtner, L., Salvaneschi, G., Freisleben, B., Mezini, M.: Fault-tolerant distributed reactive programming. In: 32nd European Conference on Object-Oriented Programming (ECOOP 2018), p. 26 (2018)
9. Ponge, J., Navarro, A., Escoffier, C., Le Mouël, F.: Analysing the performance and costs of reactive programming libraries in Java. In: Proceedings of the 8th ACM SIGPLAN International Workshop on Reactive and Event-Based Languages and Systems, pp. 51–60. ACM, Chicago IL USA (2021)
10. Hossain, T.M., Hassan, R., Amjad, M., Rahman, M.A.: Web performance analysis: an empirical analysis of e-commerce sites in Bangladesh. Int. J. Inf. Eng. Electron. Bus. **13**, 47–54 (2021)
11. Ghanavati, M., Costa, D., Seboek, J., Lo, D., Andrzejak, A.: Memory and resource leak defects and their repairs in Java projects. Empir Software Eng. **25**, 678–718 (2020)
12. Riet, J.V., Paganelli, F., Malavolta, I.: From 6.2 to 0.15 seconds – an industrial case study on mobile web performance. In: 2020 IEEE International Conference on Software Maintenance and Evolution (ICSME), pp. 746–755. IEEE, Adelaide, Australia (2020)
13. Assran, B.M., Aytekin, A., Feyzmahdavian, H.R., Johansson, M., Rabbat, M.G.: Advances in asynchronous parallel and distributed optimization. Proc. IEEE **108**, 2013–2031 (2020)
14. Sala, K., Teruel, X., Perez, J.M., Peña, A.J., Beltran, V., Labarta, J.: Integrating blocking and non-blocking MPI primitives with task-based programming models. Parallel Comput. **85**, 153–166 (2019)
15. Andre, E.: Observer patterns for real-time systems. In: 2013 18th International Conference on Engineering of Complex Computer Systems, pp. 125–134. IEEE, Singapore (2013)
16. Perez, I., Bärenz, M., Nilsson, H.: Functional reactive programming, refactored. In: Proceedings of the 9th International Symposium on Haskell. pp. 33–44. ACM, Nara Japan (2016)
17. Zhuang, X., Qing-dao-er-ji, R.: Channel congestion control model based on improved asynchronous back-pressure routing algorithm in wireless distributed networks. J. Ambient Intell. Hum. Comput. 1–11 (2020)
18. Bahr, P., Graulund, C.U., Møgelberg, R.E.: Diamonds are not forever: liveness in reactive programming with guarded recursion. Proc. ACM Program. Lang. **5**, 1–28 (2021)
19. Cheng, X., Thaeler, A., Xue, G, Chen, D.: TPS: a time-based positioning scheme for outdoor wireless sensor networks. In: IEEE INFOCOM 2004, pp. 2685–2696. IEEE, Hong Kong (2004)
20. Delasay, M., Ingolfsson, A., Kolfal, B., Schultz, K.: Load effect on service times. Eur. J. Oper. Res. **279**, 673–686 (2019)
21. Wang, Z., et al.: DeepScaling: microservices autoscaling for stable CPU utilization in large scale cloud systems. In: Proceedings of the 13th Symposium on Cloud Computing, pp. 16–30. ACM, San Francisco California (2022)
22. Heilmann, F., Fohler, G.: Size-based queuing: an approach to improve bandwidth utilization in TSN networks. SIGBED Rev. **16**, 9–14 (2019)

Improved Sparrow Search Algorithm Optimized Neural Network Analysis of Traffic Congestion

Lu Banban[ORCID] and Lian Zhigang[✉]

Shanghai Dianji University, Shanghai 201306, China
lianzg@sdju.edu.cn

Abstract. Accurate traffic congestion prediction is of great significance for applications such as traffic control and route optimization. However, the traffic situation is affected by many complex factors, and the traditional linear model is difficult to capture the nonlinear interaction information between variables. In recent years, neural networks have been widely used in the analysis and prediction of traffic congestion because of their significant advantages in identifying nonlinear and complex patterns. Firstly, aiming at the shortcomings of sparrow search algorithm, such as easy to fall into local minimum and weak global search ability, a sparrow search algorithm based on Levy flight mapping was proposed. The simulation results show that the improved algorithm can effectively overcome the limitations of the original algorithm and improve its performance in terms of convergence accuracy, stability and convergence speed. Secondly, the improved sparrow search algorithm based on Levy flight was used to find the best initial weights and thresholds of the neural network to improve the generalization ability and classification accuracy of the neural network. Finally, the optimized neural network is used to predict the state of traffic congestion. The simulation results show that the improved sparrow search algorithm can improve the performance of the neural network, and can effectively predict the future traffic congestion.

Keywords: Sparrow Search Algorithm · Levy Flight · Neural Network · Traffic Congestion · Forecasting

1 Introduction

With the rapid development of China's urbanization process and economy, the problem of urban traffic congestion is becoming increasingly serious. Coupled with the rapid increase in the number of private cars, the contradiction between road vehicles and roads has become more prominent. Traffic congestion has seriously affected the development of urban economy and the improvement of residents' living standards [1]. Therefore, how to effectively alleviate traffic congestion is a problem worth thinking about.

The traffic condition is affected by many complex factors, and the traditional linear model is difficult to identify the nonlinear information of the interaction between variables, which will cause difficulties in future condition prediction. Previous studies have used time series models, using the relationship between current and historical

J. Li et al. (Eds.): 6GN 2023, LNICST 554, pp. 109–125, 2024.
https://doi.org/10.1007/978-3-031-53404-1_10

data for prediction, and considering the periodicity and trend of the data for modeling and analysis. However, these assumptions are based on time series stability and cannot capture traffic mutation. Later, traffic prediction models based on machine learning emerged, such as k-nearest neighbor algorithm and support vector machine, which can model complex traffic flow characteristics but have limited ability to capture nonlinear patterns. In recent years, the advantages of neural networks in capturing nonlinear and complex patterns have attracted the attention of researchers, and neural networks are often used for spatio-temporal modeling in traffic forecasting at this stage.

Firstly, this paper introduces Levy flight strategy to improve the sparrow search algorithm. The Sparrow search algorithm has the defects of weak global search ability and local escape ability, and easy to fall into local minimum [2]. The Sparrow search algorithm is improved, and the simulation experiments show that compared with other algorithms, the improved algorithm has stronger local minimum escape ability and better convergence performance. Secondly, in view of the shortcomings of BP neural network, such as low learning efficiency, long training time and easy to fall into local extrema, LSSA was used to find the most suitable initial weight and threshold of the network, and the LSSA-BP neural network model was constructed [3]. Finally, the optimized neural network is used to predict the state of traffic congestion. The simulation results show that the optimized neural network has better generalization and classification accuracy, and can accurately predict the future traffic congestion.

2 Levy Flight Modified Sparrow Search Algorithm for Optimizing Neural Networks

2.1 Improved Sparrow Search Algorithm Based on Levy Flight

Levy Flight Mutation. Levy flight is a random walk, and its probability distribution of step size is heavy-tailed, which has a broader search space and stronger search ability [4]. The randomness and diversity in the search process of sparrow are enhanced by introducing Levy flight mapping. The introduction of Levy flight effectively enhances the optimization ability of the algorithm and makes the algorithm easier to jump out of local extrema [5].

The improvement mainly uses levy flight to update the population of sparrow search algorithm, that is, after the sparrow position is updated, the inertia weight factor is calculated, and the selected sparrow individual is mutated by Levy flight using roulette wheel selection. The formula is as follows:

$$X_{i,d}^{t+1} = X_{i,d}^t + \left(X_{i,d}^t - X_b^t\right) \otimes levy(d) \ if \ r > f \tag{1}$$

$X_{i,d}^t$: Represents the ith sparrow position. X_b: Is the optimal position of sparrows in the current population. f: Inertia weight factor, $f = 1 - i/MaxIter$. r: A random number of 0–1.

Finder Location Update and Improvement. The improved formula is as follows:

$$X_{i,d}^{t+1} = \begin{cases} X_{i,d}^t \cdot (1+Q) \ if \ R_2 < ST \\ X_{i,d}^t + Q \ if \ R_2 \geq ST \end{cases} \tag{2}$$

Q: Is a random number obeying normal distribution. R_2: $R_2 \in [0, 1]$, denotes the warning value. ST: ST $\in [0.5, 1]$, denotes the safe value.

In the original formula, when $R_2 < ST$, each dimension of the sparrow will gradually decrease and converge to 0, and the position update strategy in this case is not ideal. After the improved formula, the updated finder position strategy is the multiplication or addition of the current sparrow position and a random number Q, which is a normal distributed random number with mean 1 and variance 1.

Follower Location Update Improvement. The improved formula is as follows:

$$X_{i,d}^{t+1} = \begin{cases} Q \cdot exp\left(\frac{X_w^t - X_{i,d}^t}{i^2}\right) & if \ i > n/2 \\ X_b^t + \frac{\sum_{d=1}^{D} (L \cdot (|X_b^t - X_{i,d}^t|) \otimes levy(d))}{D} & other \end{cases} \tag{3}$$

X_w^t: The current worst position of the sparrow in the population. X_b^t: The current optimal sparrow position in the population. L: This value is randomly 1 or -1.

Each sparrow randomly approaches its finder. If $i \leq n/2$, the position of the sparrow is the addition and subtraction of the current best position of the sparrow with the change of the position of the sparrow with Levy flight, and evenly divided into each dimension. In this way, the deviation between one sparrow position and the current optimal position is not large, and the deviation between other positions and the current optimal position is small, and its value will converge to the current sparrow optimal position.

Alert Location Update Improvement. The improved formula is as follows:

$$X_{i,d}^{t+1} = \begin{cases} X_{i,d}^t + \beta \cdot \left(X_{i,d}^t - X_b^t\right) & if \ f_i \neq f_g \\ X_{i,d}^t + \beta \cdot \left(X_w^t - X_b^t\right) & if \ f_i = f_g \end{cases} \tag{4}$$

X_b^t: The optimal position of the sparrow in the population. β: The step size control parameter. X_w^t: The worst position of the sparrow in the population. f_g: Global optimal fitness value. f_i: The fitness value of the current sparrow.

When the sparrow is in the optimal position, it will escape between the optimal and the worst position; Otherwise, it will escape between the best position and the current position to find a more optimal solution.

Algorithm Flow

Step1: Population initialization.
Step2: Calculate the fitness value and sort it.
Step3: Sparrow updates the finder location.
Step4: Sparrow updates the follower position.
Step5: Sparrow updates the guard position.
Step6: Calculate the fitness value and update the sparrow position.
Step7: Levy flight mutation was performed on the selected sparrows.
Step8: Exit after iteration completion; Otherwise, execute Step2–7.

2.2 Simulation Experiment of Improved Sparrow Search Algorithm Based on Levy Flight

In order to verify the feasibility and effectiveness of the improved Sparrow Search Algorithm (LSSA) based on Levy flight, Particle Swarm Optimization (PSO), Sparrow Search algorithm (SSA) and Grey Wolf Algorithm (GWO) were selected as comparison algorithms. The test function is the test function of Tables 1 and 2 of Sparrow search algorithm [6]. Table 1 shows the seven unimodal test functions and Table 2 shows the five multimodal test functions with the size of 30 for each function dimension.

Table 1. Unimodal test functions

Function	Range	Best
$F_1(x) = \sum_{i=1}^{n} x_i^2$	$[-100,100]$	0
$F_2(x) = \sum_{i=1}^{n} |x_i| + \prod_{i=1}^{n} |x_i|$	$[-10,10]$	0
$F_3(x) = \sum_{i=1}^{n} \left(\sum_{j=1}^{i} x_j\right)^2$	$[-100,100]$	0
$F_4(x) = \max_i \{|x_i|, 1 \le i \le n\}$	$[-100,100]$	0
$F_5(x) = \sum_{i=1}^{n-1} \left[100\left(x_{i+1} - x_i^2\right)^2 + (x_i - 1)^2\right]$	$[-30,30]$	0
$F_6(x) = \sum_{i=1}^{n} ([x_i + 0.5])^2$	$[-100,100]$	0
$F_7(x) = \sum_{i=1}^{n} i x_i^4 + \text{random}[0, 1)$	$[-1.28,1.28]$	0

Algorithm Parameter Setting. In order to make the algorithms more convincing and realistic, the common parameters of all algorithms are kept consistent, and 30 independent trials are performed for each test function. The maximum number of iterations per trial is 500 and the population size is 50. PSO algorithm parameters setting: $C_1 = C_2 = 2$, $w = 0.9$; GWO algorithm parameter Settings: r_1, r_2 are random numbers of $[0,1]$, a decreases from 2 to 0; The parameters of the SSA and LSSA algorithms are set as follows: $PD = 0.2$, $ST = 0.8$, $SD = 0.2$. The integrated development environment of the simulation experiment is Matlab2022b and the operating system is Window11.

Analysis of Results. Table 3 (the best data has been marked in boldface) and Fig. 1 (the vertical axis of the convergence curve is treated logarithmically) show the optimization comparison results of the four algorithms on unimodal test functions.

LSSA found the optimal value on the test functions $F_1 - F_4$, and on the test functions $F_1 - F_7$, LSSA had better convergence accuracy than other algorithms. By comparing the optimal and average values, it can be seen that LSSA performs the best.

According to the standard deviation data in Table 3, the standard deviation of LSSA on $F_1 - F_4$ is 0. This indicates that LSSA is more stable than PSO, GWO and SSA algorithms. On $F_1 - F_7$ test functions, the standard deviation of LSSA is better than the other three algorithms. Simulation experiments show that LSSA has high stability and advantages compared with the other three algorithms on unimodal test functions,

Table 2. Multimodal test functions

Function	Range	Best
$F_8(x) = \sum_{i=1}^{n} -x_i \sin\left(\sqrt{\lvert x_i \rvert}\right)$	$[-500,500]$	– 418.9829n
$F_9(x) = \sum_{i=1}^{n}\left[x_i^2 - 10\cos(2\pi x_i) + 10\right]$	$[-5.12,5.12]$	0
$F_{10}(x) =$ $-20\exp\left(-0.2\sqrt{\frac{1}{n}\sum_{i=1}^{n}x_i^2}\right) - \exp\left(\frac{1}{n}\sum_{i=1}^{n}\cos(2\pi x_i)\right) + 20 + e$	$[-32,32]$	0
$F_{11}(x) = \frac{1}{4000}\sum_{i=1}^{n}x_i^2 - \prod_{i=1}^{n}\cos\left(\frac{x_i}{\sqrt{i}}\right) + 1$	$[-600,600]$	0
$F_{12}(x) =$ $\frac{\pi}{n}\left\{10\sin(\pi y_1) + \sum_{i=1}^{n-1}(y_i - 1)^2\left[1 + 10\sin^2(\pi y_{i+1})\right] + (y_n - 1)^2\right\}$ $+\sum_{i=1}^{n}u(x_i, 10, 100, 4),\, y_i = 1 + \frac{x_i+1}{4},$ $u(x_i, a, k, m) = \begin{cases} k(x_i - a)^m & x_i > a \\ 0 & -a < x_i < a \\ k(-x_i - a)^m & x_i < -a \end{cases}$	$[-50,50]$	0

indicating that the stability and robustness of LSSA are better than other algorithms, indicating that LSSA can fully and efficiently explore the search space and has strong optimization ability [7].

Figure 1 shows the convergence curves of the four algorithms. It can be observed that the convergence speed of LSSA is better than that of other algorithms on the test functions $F_1 - F_7$, and the advantage is great.

In summary, the LSSA algorithm shows a strong search ability on unimodal test functions and can quickly find the optimal value.

Table 4 and Fig. 2 show the optimization comparison results of the four algorithms on the multimodal test functions.

As can be seen from Table 4, LSSA performs better than other algorithms on the five test functions $F_8 - F_{12}$. Especially on the F_8 and F_{12} test functions, LSSA shows very good performance and finds points close to the optimal value. On the test functions F_9 and F_{11}, LSSA successfully finds the global optimal solution, and can optimize to the global minimum value each time, which proves that LSSA has powerful global search ability. On the F_9-F_{11} test function, the search ability of LSSA and SSA is equivalent. In conclusion, LSSA has a good global search ability in multimodal test functions.

On $F_8 - F_{12}$ test functions, the stability of LSSA is better than that of PSO and GWO. On the F_9-F_{11} test function, the stability of LSSA and SSA are almost the same.

Observing the convergence curve, the local extremum escape ability and convergence performance of each algorithm can be clearly compared. From the convergence curve in Fig. 2, it can be seen that the convergence accuracy of LSSA is better than that of the other three algorithms. On the test function F_9-F_{11}, LSSA shows faster convergence speed

Table 3. Comparison results of unimodal test functions

Function	Algorithm	Best	Average	STD
F_1	PSO	5.22E + 01	1.64E + 02	1.08E + 02
	GWO	1.76E-35	2.04E-33	2.46E-33
	SSA	0	1.54E-84	8.45E-84
	LSSA	**0**	**0**	**0**
F_2	PSO	3.15E + 00	9.02E + 00	5.12E + 00
	GWO	1.88E-20	8.02E-20	6.96E-20
	SSA	0	1.76E-54	9.65E-54
	LSSA	**0**	**1.79E-223**	**0**
F_3	PSO	6.90E + 02	4.68E + 03	2.93E + 03
	GWO	4.21E-11	6.97E-08	2.31E-07
	SSA	0	2.11E-142	1.04E-141
	LSSA	**0**	**0**	**0**
F_4	PSO	2.04E + 00	7.52E + 00	1.89E + 00
	GWO	5.75E-10	2.45E-08	2.85E-08
	SSA	0	5.51E-122	3.02E-121
	LSSA	**0**	**3.28E-253**	**0**
F_5	PSO	2.67E + 02	5.21E + 03	5.30E + 03
	GWO	2.60E + 01	2.68E + 01	7.68E-01
	SSA	2.35E-04	3.93E-03	5.26E-03
	LSSA	**5.95E-07**	**6.33E-04**	**1.02E-03**
F_6	PSO	5.18E + 01	1.58E + 02	8.24E + 01
	GWO	2.49E-01	5.13E-01	3.03E-01
	SSA	2.95E-07	1.01E-05	9.59E-06
	LSSA	**2.65E-11**	**2.28E-06**	**3.50E-06**
F_7	PSO	1.61E-02	3.39E-01	7.15E-01
	GWO	2.56E-04	1.30E-03	6.10E-04
	SSA	2.90E-05	3.90E-04	2.49E-04
	LSSA	**2.54E-05**	**1.32E-04**	**9.36E-05**

and needs the least number of iterations. This indicates that the introduction of Levy flight strategy contributes to the convergence speed of the algorithm. The convergence curve of LSSA algorithm shows a stepwise decline, which indicates that the introduction of Levy flight strategy can effectively avoid the algorithm falling into the local optimal solution and search for a better solution in the global scope. In contrast, the convergence curve

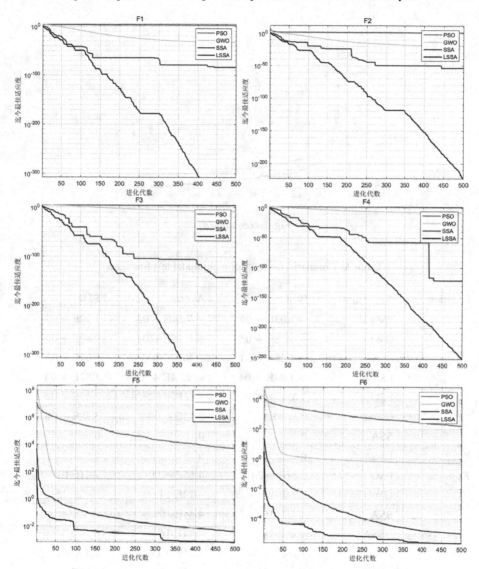

Fig. 1. Convergence curve of the unimodal test function

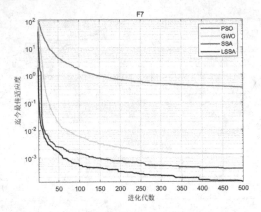

Fig. 1. (*continued*)

Table 4. Comparison results of multimodal test functions

Function	Algorithm	Best	Average	STD
F_8	PSO	− 1.03E + 04	− 7.54E + 03	1.20E + 03
	GWO	− 7.78E + 03	− 6.19E + 03	9.80E + 02
	SSA	− 1.26E + 04	− 8.10E + 03	2.04E + 03
	LSSA	**− 1.04E + 04**	**− 9.24E + 03**	**1.82E + 03**
F_9	PSO	1.05E + 02	1.88E + 02	3.55E + 01
	GWO	0	2.13E + 00	3.79E + 00
	SSA	0	0	0
	LSSA	**0**	**0**	**0**
F_{10}	PSO	2.87E + 00	4.55E + 00	7.41E-01
	GWO	3.24E-14	4.24E-14	5.63E-15
	SSA	4.44E-16	4.44E-16	0
	LSSA	**4.44E-16**	**4.44E-16**	**0**
F_{11}	PSO	1.26E + 00	2.41E + 00	7.30E-01
	GWO	0	3.13E-03	6.86E-03
	SSA	0	0	0
	LSSA	**0**	**0**	**0**
F_{12}	PSO	1.45E + 00	4.30E + 00	1.71E + 00
	GWO	6.50E-03	2.78E-02	1.42E-02
	SSA	5.82E-07	1.75E-06	1.02E-06
	LSSA	**9.78E-12**	**1.50E-07**	**2.88E-07**

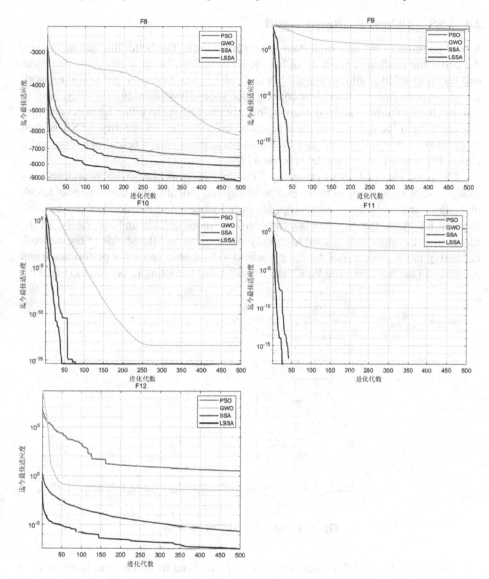

Fig. 2. Convergence curves of multimodal test functions

of traditional optimization algorithms often shows a gentle trend, and the convergence speed is relatively slow and easy to fall into stagnation.

In general, compared with other algorithms, LSSA performs well in terms of convergence accuracy, stability and convergence speed. In particular, LSSA performs well compared to different types of algorithms such as PSO and GWO. In addition, compared with similar SSA algorithms, LSSA can achieve better results and more stable solutions in most test functions. In summary, the LSSA algorithm performs well in terms of local extrema escape ability, convergence performance and stability.

2.3 LSSA-BP Neural Network Model

Model parameters are very important in neural network. The better the parameters are, the better the prediction accuracy of the model is, and the better it can fit the nonlinear and complex relationship between input and output data. However, BP neural network has some problems, such as slow convergence speed, low learning efficiency, easy to fall into local minimum and too many training times. LSSA algorithm has a good ability to escape from local minimum, convergence performance and stability. LSSA is used to optimize the initial weight and threshold parameters of BP neural network, and a LSSA-BP neural network model is formed [8].

The basic idea of LSSA optimizing BP neural network: LSSA-BP neural network model combines the local extreme escape ability and global search advantages of LSSA algorithm, and finds the most appropriate initial weight and threshold of the network through LSSA, so that the network can converge more quickly and accurately. At the same time, BP algorithm is used to modify the weights and thresholds of the network according to the direction of the error gradient to further improve the performance and accuracy of the network. The flow chart of LSSA-BP modeling is as follows (Fig. 3):

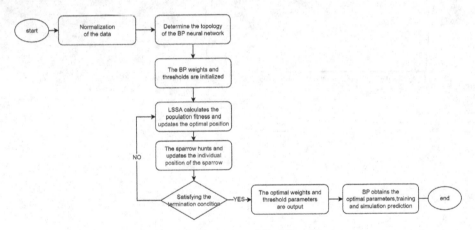

Fig. 3. LSSA-BP modeling flowchart

The hidden layer of the neural network maps the data from the input layer, and the mapping process can be expressed as follows [10]:

$$hiddenLayer\_output = F(w * x + b) \tag{5}$$

$F()$ is the activation function, w, b are the weights and thresholds, and the *hiddenLayer_output* is the output of the hidden layer for the incoming data x.

The improved sparrow search algorithm is used to optimize the neural network. The optimization direction is the process of finding the optimal model parameters (w, b) in the neural network. Therefore, the fitness function of LSSA algorithm can be designed as follows:

$$fitness = min(MSE_{TrainSet, TestSet}) \tag{6}$$

In the formula, *TrainSet*, *TestSet* are the samples of the training set and the test set, respectively.

3 LSSA-BP Neural Network Congestion Prediction

3.1 Problem Description

Traffic congestion has become a common problem in modern society, especially in developing countries, which seriously affects people's work efficiency and social development. How to effectively alleviate traffic congestion is extremely important. Accurate traffic condition prediction is helpful to traffic management and the development of urban economy.

Traffic congestion prediction is a multi-step time series prediction problem, which needs to consider the multiple influences of spatio-temporal factors and external environment, and predict the change trend of future traffic congestion conditions according to historical traffic conditions and external environment [11]. Predicting the next time step based on S time steps of historical data can be expressed as [12]:

$$\theta^* = \underset{\theta}{argmin} \sum_{t=1}^{T} \mathcal{L}_\theta \left(X^{(t-S):t}, X^{(t+1):(t+T)} \right) \tag{7}$$

$X^{(t-S):t}$: Input value, real value, here refers to the processed vehicle historical data. $X^{(t+1):(t+T)}$: The output value, the predicted value, here refers to the traffic condition of the vehicle. $\mathcal{L}_\theta$: Nonlinear loss function. θ^*: In neural networks, the optimal parameters θ^* represent the weights of the model solving the optimization problem of the nonlinear loss function $\mathcal{L}_\theta$.

In recent years, quite a few traffic prediction models have emerged, such as neural networks, which can model more complex traffic flow characteristics. However, BP neural network is easy to fall into local minimum, the network structure and parameters need to be constantly adjusted, and the training time is long, which brings difficulties to the establishment of traffic prediction model. In view of the problems existing in BP, this paper uses the improved sparrow search algorithm based on Levy flight to optimize the neural network. Then LSSA-BP is used to predict the traffic congestion in a certain area, and the simulation experiment and result analysis are carried out.

3.2 Experimental Data Processing

The data used is the GPS positioning data of taxis in Shenzhen on November 22, 2013. There are more than 40 million data in total, occupying 2.1GB of memory, including five features: car number, time, longitude, latitude, empty car (1 is passenger, 0 is empty), and speed.

Cleaning Data. The first thing you need to do with the data is remove the anomalous trajectories so that they don't affect the rest of the data.

According to the passenger status field in the data, the passenger status of a complete trajectory is either continuous 0 (no load) or continuous 1 (passenger load). If 1 or 0 suddenly appears in a trajectory, there may be data anomaly in it, and it needs to be cleaned to delete the abnormal trajectory.

Trajectory Visualization. Taxi data cannot fully represent the traffic conditions of the city, but it can select the area with dense OD (start-end point). When the sample size is large and the vehicles are dense, it can roughly represent the overall traffic conditions of the area. Figure 4 shows the OD trajectory distribution in Shenzhen.

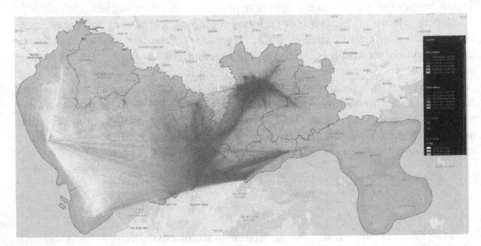

Fig. 4. Taxi OD distribution in Shenzhen

Through data visualization, it can be found that the taxis in Futian are the most dense. In order to increase the credibility of taxis to represent urban traffic conditions, the data in Futian area are used for traffic congestion prediction.

Trajectory Rasterization. The taxi trajectory data in Futian District were spatio-temporal rasterized, and the time window was set as 1200, and the spatial grid was set as 70. The average speed (aveSpeed), average acceleration (gridAcc), flow (volume), standard deviation of speed (speed_std), and average number of stops (stopNum) were calculated for each grid. The data format is shown in Fig. 5.

Determine Grid Congestion. The rasterized data does not have the corresponding congestion state label, and the road condition category in the current state of each grid is not known. Therefore, it is necessary to use a clustering method to determine the grid congestion status: congestion, slow traffic, or smooth. Using K-means algorithm [13], the samples are divided into three clusters. Two features, average speed and average number of stops, were used for clustering. Through clustering, the road condition categories (congestion, slow traffic, smooth traffic) in the current state of the grid were determined. Labels label meaning: 0 means unimpeded, 1 means congested, 2 means slow. The following figure shows the data format (Fig. 6).

Unnamed: 0	rowid	colid	time_id	aveSpeed	gridAcc	volume	speed_std	stopNum
0	0	109	-6	1.356018	-0.006263	1	0.590419	3.5
1	0	109	-5	2.820641	0.191501	1	NaN	0.0
2	0	109	-4	1.356018	-0.000514	1	0.796842	2.0
3	0	109	-3	1.356018	-0.000018	1	NaN	0.0
4	0	109	7	2.766585	0.070580	1	NaN	0.0
...	...	...	...	...	...	...	...	...

Fig. 5. Rasterized data format

Unnamed: 0	rowid	colid	time_id	aveSpeed	gridAcc	volume	speed_std	stopNum	labels
0	0	109	-6	1.356018	-0.006263	1	0.590419	3.5	1
1	0	109	-5	2.820641	0.191501	1	NaN	0.0	2
2	0	109	-4	1.356018	-0.000514	1	0.796842	2.0	1
3	0	109	-3	1.356018	-0.000018	1	NaN	0.0	2
4	0	109	7	2.766585	0.070580	1	NaN	0.0	2
...	...	...	...	...	...	...	...	...	...

Fig. 6. Data format of clustering results

3.3 Simulation Experiment of LSSA-BP Predicting Congestion Condition

Traffic Congestion Prediction Model. The change of traffic conditions is related to time. For example, the traffic conditions at time $t - 1, t - 2, t - 3$, etc. are related to the traffic conditions at time t [14]. A LSSA-BP neural network model was built to predict the traffic state of each grid for the next 10 min, each time step was 10 min, and the data was the vehicle data of the past half hour (3 time steps) to predict the next time step.

The flow chart of LSSA-BP model for traffic congestion prediction is shown in Fig. 7. The average speed, average acceleration, flow, standard deviation of speed and average number of stops of the grid are taken as the input samples of the model, and the traffic conditions (congestion, slow moving and smooth) are taken as the output samples of the model.

Parameter Setting. For the BP algorithm, the training number was set as 1000, the training goal was to achieve the minimum error of 1e-5, and the learning rate was 0.01. In LSSA algorithm, the maximum evolution generation is set as 50, the initial population size is 50, the upper and lower limits of independent variables are [−3, 3], the proportion of safe value is 0.8, the finder is 0.2, and the alert is 0.2.

Experiments and Results Analysis. The model evaluation metrics are RMSE (root mean square error), MAE (Mean absolute error), and MAPE (Mean absolute percentage error).

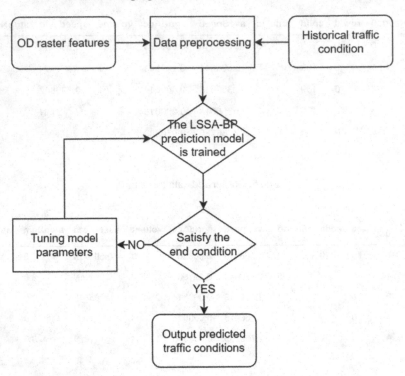

Fig. 7. Flow chart of traffic congestion condition prediction model

Figures 8 and 9 shows the experimental results of fitting the vehicle data into the LSSA-BP model. Figure 8 show the comparison between the predicted value and the real value of LSSA-BP and BP. In contrast, the predicted value of BP and the real value deviate greatly, and the predicted result of LSSA-BP is more accurate. Figure 9 are error comparison plots, which show that the prediction deviation of BP is relatively large on both the training set and the test set, and the error of LSSA-BP is significantly smaller than that of BP.

Table 5 shows the parameters of the experimental results. In contrast, the RMSE, MAE and MAPE of LSSA-BP are significantly smaller than those of BP. The smaller error indicates that the predicted data and the original data are more overlapped, and the prediction accuracy of the model is better, that is, LSSA-BP can predict the future traffic congestion more accurately.

The experimental results show that compared with the traditional BP algorithm, LSSA-BP has better performance and better prediction accuracy, which indicates that the LSSA algorithm can improve the convergence speed and learning efficiency of BP neural network, thereby improving its performance in practical applications.

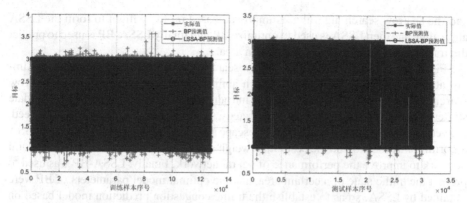

Fig. 8. Plot of predicted versus true values

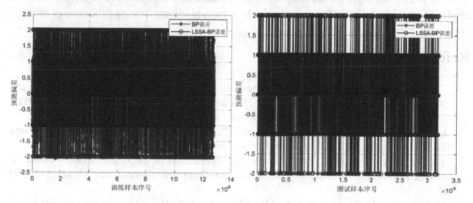

Fig. 9. Prediction error plot

Table 5. Error parameter

Training set	RMSE	MAE	MAPE
BP	0.43	0.28	16.02%
LSSA-BP	**0.34**	**0.08**	**4.02%**
Test set	RMSE	MAE	MAPE
BP	0.43	0.29	17.21%
LSSA-BP	**0.33**	**0.07**	**4.19%**

4 Conclusion

Traffic congestion has caused a series of problems, which not only hinder the development of urban economy and transportation, but also seriously affect the convenience of citizens' lives. Accurate traffic congestion prediction is helpful to improve traffic efficiency [15]. In this paper, LSSA-BP is used to predict the traffic congestion situation, and the sparrow search algorithm is improved based on Levy flight to form the LSSA algorithm, and then LSSA is used to optimize BP, and finally LSSA-BP is used to predict the traffic congestion situation in a certain area.

Firstly, this paper improves the sparrow search algorithm based on Levy flight. Aiming at the problem of weak global search ability and local escape ability of Sparrow search algorithm, Levy flight strategy is introduced to optimize the population update strategy and individual iterative update strategy, so as to improve the convergence speed, convergence accuracy and local extreme escape ability of the algorithm. Then, the LSSA algorithm was used to find the best initial weights and threshold parameters of BP neural network to improve the performance of neural network. Finally, LSSA-BP was used to predict the traffic state of a certain area, and the optimal model parameters of BP were obtained by LSSA, so as to establish the traffic congestion prediction model based on LSSA-BP. The simulation results show that LSSA-BP can predict the future traffic congestion well. And with the development of artificial intelligence, the algorithm can be applied to a wider range of scenarios, such as path planning, image processing, etc.

References

1. Zhang Mingjie, F., Wu Jianhong, S.: Research on Traffic congestion in Xi'an City from the perspective of information management. J. Xi'an Univ. Posts Telecommun. **17**(01), 114–117 (2012)
2. Li Yali, F., Wang Shuqin, S., Chen Qianru, T.: Comparative study of several new swarm intelligence optimization algorithms. Comput. Eng. Appl. **56**(22), 1–12 (2020)
3. Yan Xu, F., Li Siyuan, S., Zhang Zheng, T.: Application of BP neural network based on genetic algorithm in prediction of urban water consumption. Comput. Sci. **43**(S2), 547–550 (2016)
4. Mao Qinghua, F., Zhang Qiang, S., Mao Chengcheng, T.: Hybrid sine cosine algorithm and levy flight sparrow algorithm. J. Shanxi Univ. **44**(06), 1086–1091 (2021)
5. Liu Ziyang, F., Pang Zhihua, S., Tao Pei, T.: Memory-enhanced levy flight gravitational search algorithm. Comput. Simul. **39**(01), 312–317 (2022)
6. Xue, J., Shen, B.: A novel swarm intelligence optimization approach: sparrow search algorithm. Syst. Sci. Control Eng. **8**(1), 22–34 (2020)
7. Fu Hua, F., Liu Hao, S.: Improved sparrow search algorithm with multi-strategy fusion and its application. Control Decis. **37**(01), 87–96 (2022)
8. Liu Yuan, F., Wang Fang, S.: Sparrow search algorithm optimized BP neural network for short-term wind power prediction. J. Shanghai Inst. Electr. Technol. **25**(03), 132–136 (2022)
9. Zhou Yi, F., Hu Shuting, S., Li Wei, T.: Traffic prediction technology driven by graph neural network: exploration and challenges. J. Internet Things **5**(4), 1–16 (2021)
10. Liu Yong, F., Zhang Liyi, S.: Implementation and performance comparison of BP and RBF neural networks. Electron. Measur. Technol. **30**(4), 77–80 (2007)
11. Kong, X., Zhang, J., Wei, X., et al.: Adaptive spatial-temporal graph attention networks for traffic flow forecasting. Appl. Intell. **2**, 1–17 (2021)
12. Bui, K.-H.N., Cho, J., Yi, H.: Spatial-temporal graph neural network for traffic forecasting: an overview and open research issues. Appl. Intell. **52**(3), 2763–2774 (2022). https://doi.org/10.1007/s10489-021-02587-w
13. Yang Junchuang, F., Zhao Chao, S.: A survey on k-means clustering algorithm. Comput. Eng. Appl. **55**(23), 7–14 (2019)

14. Gao, Y., Zhou, C., Rong, J., Wang, Y., Liu, S.: Short-term traffic speed forecasting using a deep learning method based on multitemporal traffic flow volume. IEEE Access **10**, 82384–82395 (2022). https://doi.org/10.1109/ACCESS.2022.3195353
15. Yang Xinru, F.: Research on solving the problem of urban road traffic congestion. Sci. Technol. Inf. **5**, 204 (2010)

Industrial Noisy Speech Enhancement Using Joint Time-Frequency Loss Function Based on U-Net

Rongxin Qin[ID] and Zhigang Lian[✉]

The School of Electronic and Information Engineering, Shanghai Dianji University,
Shanghai 201306, China
lianzg@sdju.edu.cn

Abstract. Single-channel speech enhancement research in complex industrial production environments is limited. Current methods, whether based on attention mechanisms or generative adversarial networks, primarily focus on learning speech characteristics in the time domain, neglecting the frequency spectrum. Additionally, existing frequency-domain algorithms lack accuracy in spectral and phase matching of noisy speech, rendering them unsatisfactory for industrial noise environments. To address this issue, this paper proposes the TFU-Net model, a time-frequency joint loss function algorithm based on deep learning and U-Net. It incorporates a combined loss function of Least Absolute Error (LAE) and Mean Square Error (MSE) for speech enhancement in industrial noise environments. Experimental results demonstrate that the frequency-domain loss function, when combined with the time-domain loss function, yields better speech enhancement under industrial environmental noise.

Keywords: Speech enhancement · Joint time-frequency loss · Deep learning

1 Introduction

Voice enhancement is used to remove background or human noise when speaking, improve speech quality, and make speech more understandable. Research into speech enhancement in recent years has focused on how to eliminate interference noise frequency without destroying the audio containing useful information. Speech enhancement is an important part of speech signal processing. Due to the influence of noise and human factors, it is not easy to obtain clean speech. Therefore, a speech enhancement algorithm is needed to recover clean speech from noisy speech as much as possible.

Depending on the number of microphones used to collect the voice, speech enhancement can be divided into single-channel and multi-channel speech

J. Li et al. (Eds.): 6GN 2023, LNICST 554, pp. 126–136, 2024.
https://doi.org/10.1007/978-3-031-53404-1_11

enhancement. As only one voice signal is collected, the correlation between multiple voice signals cannot be utilized, making noise reduction more difficult. Traditional single-channel speech enhancement methods include Wiener filtering [1], spectral subtraction [8], and minimum mean square error estimation [4]. These methods assume that the noise is stationary, but in reality, it is usually non-stationary noise. Traditional methods do not effectively remove non-stationary noise. In recent years, with the continuous development of deep learning methods, convolutional neural networks (CNN) [2,6,7], deep neural networks (DNN) [9,10,14], and long short-term memory (LSTM) [5,9,11] networks with strong fitting abilities have opened up new possibilities for speech enhancement research. Xu Yong et al. [13] take the clean speech logarithmic power spectrum as the target, constructing the mapping function between the logarithmic power spectra (LPS) of noisy speech and the clean speech logarithmic power spectrum through the training network, to improve the speech enhancement effect in non-stationary noise. Reference [3] For single-channel speech enhancement, a noise-aware attention-gated network (NAAGN) is proposed, which integrates deep residual learning, extended convolution, and attention-gating mechanisms into a U-Net architecture, expanding the receiving domain to systematically aggregate contextual information, and further improving prediction accuracy. A new noise perception multitasks loss function, the weighted average absolute error loss function is presented. The NAAGN method is experimentally validated to improve the segmented signal-to-noise ratio, speech quality, and clarity. Reference [12] presents a recurrent neural network (RNN) model with four layers of LSTM hidden layers. Their experimental results show that the RNN-based model performs better than the DNN-based model for speech enhancement.

2 Joint Time-Frequency Convolutional Network

2.1 Convolutional Network Architecture

This paper introduces a convolutional neural network (CNN) model called TFU-Net, which is based on U-Net and incorporates a customized spatiotemporal joint loss function. The model is constructed using the TensorFlow 2.0 framework, as illustrated in Fig. 1.

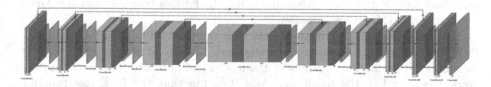

Fig. 1. TFU-Net Convolutional neural network model

The TFU-Net model adopts an encoder-decoder architecture, incorporating batch normalization (BN) layers to accelerate neural network training, improve model generalization, and reduce overfitting. The encoder consists of a series of convolutional and max-pooling layers, gradually reducing the spatial dimensions of the input feature maps. The decoder, on the other hand, consists of a series of upsampling layers and convolutional layers, increasing the spatial dimensions of the feature maps to generate the final output. The encoder and decoder are connected through skip connections, allowing the decoder to access high-resolution features from the encoder. The TFU-Net encoder comprises four levels, each consisting of two convolutional layers and one max-pooling layer. The convolutional layers in the first two layers of each level have a filter size of 3×3, 'same' padding, and a ReLU activation function. The max-pooling layer has a pooling size of 2×2 with a stride of 2. The number of filters in the convolutional layers increases from 32 to 256 as the spatial dimensions decrease. Corresponding to the encoder, the TFU-Net decoder also has four levels, each consisting of an upsampling layer and two convolutional layers. The upsampling layer has a size of 2×2 with a stride of 2, and the convolutional layers have a filter size of 3×3, 'same' padding, and ReLU activation function. The number of filters in the convolutional layers decreases from 256 to 32 as the spatial dimensions increase. In TFU-Net, skip connections connect the encoder and decoder at corresponding levels. The output feature maps from the encoder levels are connected to the upsampled decoder feature maps before the convolutional layers. This allows the decoder to access high-resolution features from the encoder, aiding in the preservation of spatial details in the final output.

TFU-Net employs a customized spatiotemporal joint loss function for model training, combining the LAE loss function and the MSE loss function in the time and frequency domains. The LAE loss function is used to enhance the detailed information of signal features, while the MSE loss function penalizes significant errors in the predicted time-frequency representation. The joint loss function ensures accurate reconstruction of the temporal and spectral components of the input in the output.

2.2   Loss Function

The loss function of a neural network model is a function that measures the difference between the model's predicted results and the ground truth values. In applications involving the removal of industrial noise, commonly used loss functions include the Least Absolute Error (LAE) loss function and the Mean Squared Error (MSE) loss function.

The LAE loss function quantifies the time-domain and frequency-domain distance between the enhanced speech signal and the original clean speech signal. It measures the prediction accuracy of the model and optimizes the model parameters. Compared to the traditional MSE loss function, the LAE loss function provides a more balanced penalty for prediction errors, avoiding dominance by extreme values. Additionally, it exhibits better performance in handling non-Gaussian distributed noise.

The MSE loss function calculates the mean of the squared differences between predicted values and ground truth values. Its advantage lies in penalizing outliers, effectively reducing errors. However, its sensitivity to outliers can make it susceptible to noise, leading to overfitting of the output results.

In applications involving the removal of industrial noise, it is often necessary to consider both the accuracy and robustness of the model while reducing sensitivity to noise. Therefore, a joint loss function that combines LAE and MSE, such as the customized spatiotemporal joint loss function mentioned in this paper, can be employed. This type of joint loss function preserves both time-domain and frequency-domain information and exhibits a certain tolerance towards outliers. As a result, it can better address the task of industrial noise removal.

The Least Absolute Error (LAE) Loss Function. In the field of signal processing, the LAE loss function is commonly used in autoregressive models. It quantifies the time-domain and frequency-domain distance between the enhanced speech signal and the original speech signal to measure the model's prediction accuracy and optimize model parameters. The LAE loss function helps the model preserve fine details more effectively.

1)The expression for the temporal LAE loss function is as follows:

$$T_1 = \frac{1}{k} \sum_{m=1}^{k} |\hat{a}_m - a_m| \tag{1}$$

In the equation, k represents the number of sampled points in the speech signal, $\hat{a}_m$ represents the predicted value of the m-th sampled point in the enhanced speech signal, and a_m represents the original value of the m-th sampled point in the original speech signal.

2)The expression for the frequency-domain LAE loss function is as follows:

$$F_1 = \frac{1}{W} \sum_{m=1}^{W} ||STFT(a_m)| - |STFT(\hat{a}_m)|| \tag{2}$$

In the equation, W represents the number of frames in the speech signal, $|STFT(a_m)|$ represents the magnitude of the m-th frame of the original speech amplitude spectrum after Fast Fourier Transform (FFT), and $|STFT(\hat{a}_m)|$ represents the magnitude of the m-th frame of the enhanced speech amplitude spectrum after FFT.

The Mean Squared Error (MSE) Loss Function. The MSE loss function is a common loss function used to measure the average difference between model predictions and ground truth values. It is often employed in the training of speech enhancement models. In the field of speech enhancement, the MSE loss function is utilized to quantify the difference between the output speech and the target speech, enabling the model to better learn how to suppress noise.

1)The expression for the temporal MSE loss function is as follows:

$$T_2 = \frac{1}{k} \sum_{m=1}^{k} (a_m - \hat{a}_m)^2 \tag{3}$$

In the equation, k represents the number of sampled points in the speech signal, a_m represents the original value of the m-th sampled point in the original speech signal, and $\hat{a}_m$ represents the predicted value of the m-th sampled point in the enhanced speech signal.

2)The expression for the frequency-domain MSE loss function is as follows:

$$F_2 = \frac{1}{W} \sum_{m=1}^{W} (|STFT(a_m)| - |STFT(\hat{a}_m)|)^2 \qquad (4)$$

In the equation, W represents the number of frames in the speech signal, $|STFT(a_m)|$ represents the magnitude of the m-th frame of the original speech amplitude spectrum after FFT, and $|STFT(\hat{a}_m)|$ represents the magnitude of the m-th frame of the enhanced speech amplitude spectrum after FFT.

Joint Time-Frequency Loss Function. In speech enhancement tasks, although the LAE loss function can be applied, it does not completely match the time-domain and frequency-domain characteristics of speech. The MSE loss function also has some drawbacks, such as sensitivity to outliers, susceptibility to noise and distortion, and an inability to preserve local details and structural information of the speech signal effectively. Therefore, this paper introduces the MSE loss function to complement the LAE loss function in training, aiming to better adapt to the training objective of speech enhancement. Additionally, leveraging the ability of the LAE loss function to preserve fine details, compensates for the limitations of the MSE loss function. Building upon the aforementioned LAE/MSE loss functions in their time-domain and frequency-domain forms, four spatiotemporal joint loss functions are constructed, as shown below.

$$L_1 = \varepsilon \cdot \frac{1}{k} \sum_{m=1}^{k} |\hat{a}_m - a_m| + (1-\varepsilon) \cdot \frac{1}{W} \sum_{m=1}^{W} \||STFT(a_m)| - |STFT(\hat{a}_m)\| \quad (5)$$

$$L_2 = \varepsilon \cdot \frac{1}{k} \sum_{m=1}^{k} |\hat{a}_m - a_m| + (1-\varepsilon) \cdot \frac{1}{W} \sum_{m=1}^{W} (|STFT(a_m)| - |STFT(\hat{a}_m)|)^2 \quad (6)$$

$$L_3 = \varepsilon \cdot \frac{1}{k} \sum_{m=1}^{k} (a_m - \hat{a}_m)^2 + (1-\varepsilon) \cdot \frac{1}{W} \sum_{m=1}^{W} \||STFT(a_m)| - |STFT(\hat{a}_m)\| \quad (7)$$

$$L_4 = \varepsilon \cdot \frac{1}{k} \sum_{m=1}^{k} (a_m - \hat{a}_m)^2 + (1-\varepsilon) \cdot \frac{1}{W} \sum_{m=1}^{W} (|STFT(a_m)| - |STFT(\hat{a}_m)|)^2 \quad (8)$$

3   Experiments

3.1   Experiment Environment and Dataset

The experimental system environment in this paper is 64-bit Ubuntu 20.04 LTS. The primary open-source tool used is TensorFlow 2.0. For model training and testing, the AISHELL-1 Chinese speech dataset and the ESC-50 open-source noise dataset were utilized.

In terms of dataset design, this experiment focuses on 9 categories of industrial environmental noise, which are: knocking sounds, chainsaw sounds, footsteps, car horn sounds, glass breaking sounds, breathing sounds, coughing sounds, keyboard typing sounds, and machine engine sounds. Subsequently, these data are processed with a sampling rate of 16kHz and a window length of at least 1 s. The noise is randomly mixed with clean speech, resulting in a training dataset of approximately 13 h of noisy speech and a validation dataset of approximately 2 h of noisy speech.

3.2 Experimental Design

In terms of model parameters, the TFU-Net has a data sample size of 100,000, a training batch size of 100, a minimum speech interval of 1 s, and a window interval of 1 s for training data. The Adam optimizer is used with a learning rate of 0.0001. The evaluation metrics for the model are Perceptual Evaluation of Speech Quality(PESQ) and Short-Time Objective Intelligibility(STOI).

The structural parameters of the constructed TFU-Net convolutional neural network model are presented in the following (Table 1): In the experiment, let it be known that S=12, indicating that both the downsampling and upsampling operations are conducted 12 times. The TFU-Net model employs an encoder-decoder architecture, progressively reducing the spatial dimension of feature maps to extract abstract representations, followed by upsampling in the decoder to

Table 1. Structural Parameters of TFU-Net Convolutional Neural Network Model

S	Encoder			Decoder		
	Input Size	Output Size	Params	Input Size	Output Size	Params
1	(128,128,1)	(128,128,16)	160	(8,8,256)	(16,16,128)	131200
2	(128,128,1)	(128,128,16)	1160	(16,16,128)	(16,16,128)	295040
3	(128,128,16)	(128,128,16)	2320	(16,16,128)	(16,16,128)	147584
4	(64,64,16)	(64,64,32)	4640	(16,16,128)	(32,32,64)	32832
5	(64,64,32)	(64,64,32)	9248	(32,32,64)	(32,32,64)	73792
6	(64,64,32)	(32,32,64)	18496	(32,32,64)	(32,32,64)	36928
7	(32,32,64)	(32,32,64)	36928	(32,32,64)	(64,64,32)	8224
8	(32,32,64)	(16,16,128)	73856	(64,64,32)	(64,64,32)	18464
9	(16,16,128)	(16,16,128)	147584	(64,64,32)	(64,64,32)	4128
10	(16,16,128)	(8,8,256)	295168	(128,128,32)	(128,128,16)	4624
11	(8,8,256)	(8,8,256)	590080	(128,128,16)	(128,128,16)	4624
12	(8,8,256)	(8,8,256)	590080	(128,128,16)	(128,128,2)	290
Output				Tanh		
				(128,128,2)	(128,128,1)	3

gradually restore the spatial resolution of the feature maps. Furthermore, the model incorporates batch normalization (BN) layers to expedite training and enhance generalization capabilities, as well as skip connections that link the encoder and decoder, thereby preserving high-resolution feature information within the decoder. The encoder and decoder of TFU-Net consist of four levels each, where the convolutional layers utilize 3×3 filter kernels, and the maximum pooling layers employ a pooling size of 2×2. Notably, the number of filters in the encoder's convolutional layers progressively increases, while it decreases in the decoder's convolutional layers. The skip connections connect corresponding levels of the encoder and decoder, facilitating the concatenation of output feature maps from the encoder level with upsampled feature maps from the decoder level. In the TFU-Net model, a tanh layer is employed as the final layer of the decoder, serving to map the linear output of the decoder to the range of $[-1, 1]$. This approach effectively limits the dynamic range of the output signal within $[-1, 1]$, thereby mitigating distortion and noise. Additionally, the tanh layer promotes signal smoothness, reducing high-frequency noise in the output signal. Thus, the inclusion of the tanh layer enhances the performance and stability of the TFU-Net model. Such a structure of TFU-Net contributes to improved expressive capacity and enables the learning of deep-level features, resulting in superior performance for speech enhancement tasks involving industrial noise reduction.

The application scenario targeted by this experiment involves speech enhancement in industrial noise environments. The evaluation primarily revolves around analyzing the frequency component distribution, PESQ, and STOI of the spectrograms and waveforms before and after denoising. Accordingly, the experimental study in this paper is divided into the following two groups.

1)The selection of weight parameter ϵ and the impact of different loss functions on the performance of the TFU-Net network are examined in this experiment. Given that the LAE loss function and the MSE loss function complement each other, the magnitude of their respective weights significantly influences model training. This experiment compares the effects of different time-frequency loss functions under various values on the model's training performance. Consequently, the experiment results in the identification of the optimal values and the most suitable loss function based on the dataset used in this study.

2)The evaluation focuses on assessing the enhancement effect of the trained model on noisy speech data. The background noise utilized in this experiment is derived from a factory's real-world environment. The recordings of speakers are conducted within this noise environment at a sampling rate of 16,000. The experimental results are visually presented through spectrograms and power spectra of the speech signals, offering an intuitive representation of the outcomes.

3.3   Results

1)The selection of weight parameter ϵ and the impact of different loss functions on the performance of the TFU-Net network.

As indicated in the table below, when epochs=30, a comparison of the training and validation losses for different time-frequency loss functions under

Table 2. The training loss and validation loss of the model under different ϵ values for various time-frequency loss functions.

	L1		L2		L3		L4	
	train loss	val loss	train loss	val loss	train loss	val loss	train loss	val loss
$\epsilon=0.1$	0.0062	0.0055	0.0128	0.0131	0.0177	0.0179	0.0083	0.0078
$\epsilon=0.3$	0.0039	0.0036	0.0141	0.0143	0.0161	0.0165	0.0067	0.0061
$\epsilon=0.5$	0.0028	0.0025	0.0136	0.0129	0.0149	0.0144	0.0053	0.0049
$\epsilon=0.7$	0.0041	0.0056	0.0163	0.0168	0.0136	0.0151	0.0075	0.0083
$\epsilon=0.9$	0.0076	0.0051	0.0187	0.0184	0.0119	0.0107	0.0087	0.0079

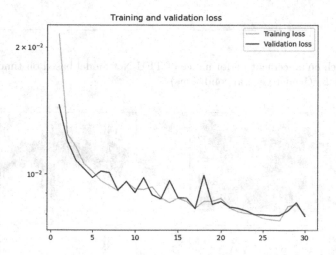

Fig. 2. When ϵ was 0.5, the training and validation loss of loss function L_1

varying ϵ values during model training was conducted. Based on this experiment, the relative optimal value for c and the best loss function was determined. From Table 2 and Fig. 2, it can be observed that with an ϵ value of 0.5 and using Function L_1 as the loss function, the model achieves the lowest training loss and validation loss, indicating that it performs the best among the tested options.

2)Evaluating the enhancement effect of the trained model on noisy speech data.

Table 3 presents the PESQ and STOI scores for denoised speech using TFU-Net with a time-frequency joint loss function and U-Net with a time-domain loss function. The speech samples used in the evaluation contained factory noise.

Table 3. Comparison of denoised effect between TFU-Net and U-Net

	PESQ	STOI
TFU-Net	1.2640	0.7587
U-Net	0.9227	0.1360

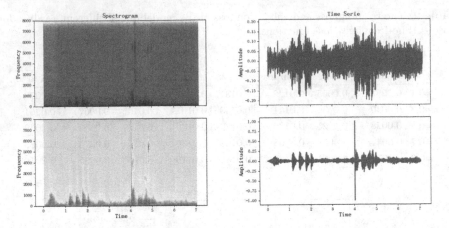

Fig. 3. Speech enhancement performance of TFU-Net model based on time-frequency loss function L_1 (Under general conditions)

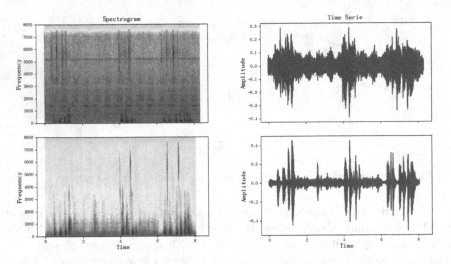

Fig. 4. Speech enhancement performance of TFU-Net model based on time-frequency loss function L_1(Under extreme conditions)

Based on Table 3, Figs. 3 and 4, it can be observed that under both typical and extreme factory noise conditions, the model exhibits effective removal of mid-to-high frequency noise, thus confirming the efficacy of the proposed convolutional neural network model based on U-Net architecture and the utilization of a custom time-frequency joint loss function for speech enhancement in factory noise scenarios.

4   Conclusion

In the experiment, a substantial amount of testing and comparison was conducted using a real-world speech dataset from an industrial production environment. The experimental results indicate that incorporating a frequency domain loss function yields better adjustment of the weight magnitude of the time domain loss function, resulting in improved speech enhancement compared to traditional time-domain loss-based algorithms. Additionally, this study optimized the design of the convolutional neural network by employing deeper convolutional layers and a greater number of filters, thereby enhancing the complexity and feature extraction capabilities of the network. The speech enhancement algorithm based on convolutional neural networks and time-frequency joint loss function represents a promising research direction. In the future, further optimization of algorithm design and parameter adjustment can enable its application in a wider range of industrial production scenarios. Moreover, exploring additional deep learning models and loss functions can continuously enhance the effectiveness of speech enhancement. The achievements of these efforts are expected to provide strong support for applications such as intelligent manufacturing and human-machine interaction, thereby driving the advancement of speech signal processing technology.

References

1. Boll, S.: Suppression of acoustic noise in speech using spectral subtraction. IEEE Trans. Acoust. Speech Signal Process. **27**(2), 113–120 (1979)
2. Chua, L.O., Roska, T.: The CNN paradigm. IEEE Trans. Circ. Syst. I Fundam. Theory Appl. **40**(3), 147–156 (1993)
3. Deng, F., Jiang, T., Wang, X., Zhang, C., Li, Y.: NAAGN: noise-aware attention-gated network for speech enhancement. In: Interspeech, pp. 2457–2461 (2020)
4. Ephraim, Y., Malah, D.: Speech enhancement using a minimum-mean square error short-time spectral amplitude estimator. IEEE Trans. Acoust. Speech Signal Process. **32**(6), 1109–1121 (1984)
5. Greff, K., Srivastava, R.K., Koutník, J., Steunebrink, B.R., Schmidhuber, J.: LSTM: a search space odyssey. IEEE Trans. Neural Networks Learn. Syst. **28**(10), 2222–2232 (2016)
6. He, K., Gkioxari, G., Dollár, P., Girshick, R.: Mask R-CNN. In: Proceedings of the IEEE International Conference on Computer Vision, pp. 2961–2969 (2017)
7. Kattenborn, T., Leitloff, J., Schiefer, F., Hinz, S.: Review on convolutional neural networks (CNN) in vegetation remote sensing. ISPRS J. Photogramm. Remote. Sens. **173**, 24–49 (2021)
8. Lim, J., Oppenheim, A.: All-pole modeling of degraded speech. IEEE Trans. Acoust. Speech Signal Process. **26**(3), 197–210 (1978)
9. Mittal, S.: A survey on modeling and improving reliability of DNN algorithms and accelerators. J. Syst. Architect. **104**, 101689 (2020)
10. Pan, J., Liu, C., Wang, Z., Hu, Y., Jiang, H.: Investigation of deep neural networks (DNN) for large vocabulary continuous speech recognition: Why DNN surpasses GMMs in acoustic modeling. In: 2012 8th International Symposium on Chinese Spoken Language Processing, pp. 301–305. IEEE (2012)

11. Staudemeyer, R.C., Morris, E.R.: Understanding LSTM-a tutorial into long short-term memory recurrent neural networks. arXiv preprint arXiv:1909.09586 (2019)
12. Takeuchi, D., Yatabe, K., Koizumi, Y., Oikawa, Y., Harada, N.: Real-time speech enhancement using equilibriated RNN. In: ICASSP 2020–2020 IEEE International Conference on Acoustics, Speech and Signal Processing (ICASSP), pp. 851–855. IEEE (2020)
13. Xiang, Y., Bao, C.: Speech enhancement via generative adversarial LSTM networks. In: 2018 16th International Workshop on Acoustic Signal Enhancement (IWAENC), pp. 46–50. IEEE (2018)
14. Zhang, J., Zheng, Y., Qi, D., Li, R., Yi, X.: DNN-based prediction model for spatio-temporal data. In: Proceedings of the 24th ACM SIGSPATIAL international conference on advances in geographic information systems, pp. 1–4 (2016)

Multiple Color Feature and Contextual Attention Mechanism Based on YOLOX

Shuaidi Shan, Pengpeng Zhang$^{(\boxtimes)}$, Xinlei Wang, Shangxian Teng, and Yichen Luo

Shanghai Dianji University, Shuihua Road 300, Shanghai, China
zhangpp@sdju.edu.cn

Abstract. Object detection aims to find out and classify objects in which people are interested. YOLOX is the one-stage object detector representative, with being famous for its quick speed. Nevertheless, recent studies have illustrated that YOLOX suffers from small-scale accuracy. To deal with this issue, enlarging recent datasets and training models are adopted. It, however, results apply effectively on those big datasets, not on small or personal self-collected ones. It means a lack of datasets on special situations, like plastic runway surfaces. To enhance the performance on detecting small-scale objects, in this paper, we proposed Double-C YOLOX, an improved algorithm based on YOLOX. The model adds an HSV module and convolutional block attention to achieve more feature extraction. Due to the scarcity of plastic runway surface and the similarity of road damage, the published road dataset and selfie dataset are combined to train and test the performance of the proposed method. Experiments show that our model improves the mAP score by 2.82%. Double-C YOLOX is more suitable for detecting the small damages, such as hole and crack, than YOLOX.

Keywords: Object detection · YOLOX · Small-scale object

1 Introduction

Object detection is one of the main parts of computer vision, originating and generation from other two field: object location and object classification. It is proven to be fitting for industrial situations to cut the cost of humans. In 2012 ImageNet Large Scale Visual Recognition Challenge (ILSVRC) [1], Hinton and Krizhevsky [2] had applied CNN to image classification and showed its wonderful performance, decreasing the rate of error from 26.2% to 15.3%. In short, there is the possibility of applying this automated approach to real scenario with improved accuracy.

This work is supported by National Natural Science Foundation of China Youth Fund (No. 61802247), Natural Science Foundation of Shanghai (No. 22ZR1425300) and Other projects of Shanghai Science and Technology) Commission (No. 21010501000). "Plastic Runway Surface Damage Detection System with YOLO" project for the Shanghai 2022 College Student Innovation Competition.

J. Li et al. (Eds.): 6GN 2023, LNICST 554, pp. 137–148, 2024.
https://doi.org/10.1007/978-3-031-53404-1_12

Recently, deep learning-based object detection algorithms developed as soon as possible. There are two branches. One is two-stages head by the ROI (Region Of Interest)-pooling layer of Fast R-CNN [3], and another is one-stage head by obtaining the region of interest very fast YOLO [4] series. The series of YOLO replace the concept of generating candidate boxes, improving its speed and reaching at 45fps. Admittedly, this method provides a fairly nice structure that balances the issues of speed and accuracy. There will be two problems with respect to this approach. Firstly, the accuracy of detecting small-scale object is worse. Poor performance of small object detection is mainly owing to the limitations of the structure of the model itself and the characteristics of small objects themselves [5]. To solve this problem, Chen et al. [6] scaled and stitched images to achieve better enhancement effects. Although this is efficient, it limits to apply of pre-training. Secondly, the general models cannot be applied directly to resolve practical questions in special area of the real world.

In this paper, we aim to design a system, focusing on abstract color features and convolutional block attention (Double-C). In our system, a small object detection algorithm based on YOLOX [7] with a color module and a CBAM [8] module is proposed to abstract more contextual information before the Neck structure. Furthermore, the small data set is addressed using two published roadway damage datasets. Because we found that the cause of damages, why it occurs, is similar in both two fields. The proposed method achieves an increase in model accuracy, which can detect more small-scale objects in the field of plastic surface damage.

To the best of our knowledge, none of the prior research has explored the area of plastic runway surfaces. Our model has verified the importance of luminance on the overall performance of the model when extracting shallow features.

The contributions of this paper are twofold:

1) We propose a method of abstracting color feature, helping the model preserve shallow features.
2) We have verified the usability of this model in the field of plastic racetracks, and, to some extent, fill the gap in this area.

2 Proposed Model

YOLOX is a high-performance object detection model that aims to resolve the problem of balance accuracy and speed. The series of YOLO is based on an object regression problem. It means that the detection process involves dividing the input image into a certain grid according to certain specifications, and traversing the entire image once. The structure of YOLOX consists of three main parts, including the Backbone network, neck network, and head network. The backbone network named CSPDarknet53 refers to the backbone network design in YOLOv3 [9], but without a fully connected layer. YOLOX uses the Spatial Pyramid Pooling (SPP) and feature pyramid network (FPN) as its neck network. Through building FPN, the model fuses the feature maps from different levels. Also, the model enhances the spatial resolution of the feature maps, which has been shown to improve detection accuracy and speed. In addition, the head network is responsible for predicting the object locations and categories based on the feature maps. Overall, YOLOX achieves quite outstanding performance on several benchmark datasets, while maintaining a fast inference speed at that time.

Although the YOLOX model shows better performance in identifying large-scale objects, it presents limitations in detecting small-scale objects. One reason is that small ones have little feature information from the relevant context. In addition, for the model of the YOLO algorithm itself, the boxing size of medium and small targets is fixed. After choosing YOLOX-S to train self-collected data sets, the model makes some medium and large targets easy to be divided into multiple local and single small targets. To solve these problems, we coin a new small object detection algorithm (Double-C YOLOX), applied to detect two defects: cracks and holes in plastic runway surfaces. The main structure refers to YOLOX, adds an HSV module in pre-training and CBAM (Convolutional Block Attention Module) after the backbone network in each layer. We formalize this by proposing a simple network as shown in Fig. 1. The Double-C YOLOX is embed the HSV module to find the color difference from the specific area, And the CBAM module pay more attention to contextual information. Both modules related concepts and reasons for use will introduce in the next 2.1 and 2.2. Both help the new model obtain richer small objects' feature information, improving the model accuracy effectively.

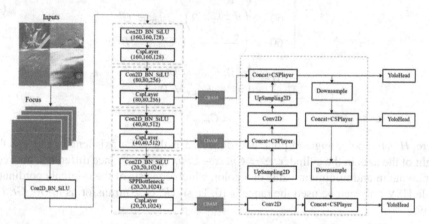

Fig. 1. The illustration of Double-C Model. It shows some improvements to the YOLOX algorithm. The input image has been preprocessed by HSV module and Mosaic at first. And three locations are used for embedding the CBAM module.

2.1 HSV Module

Before introducing how to use HSV module, there are some basic concepts. Two types of color space are popular that one is HSV, another is RGB. HSV color space are designed to capture the color features of the input images. But there are several differences. RGB color space is comprised of Red, Green, and Blue, while HSV includes Hue, Saturation, and Value. Although both can abstract color features from the input images, HSV space is more intuitive for color selection and manipulation, because it maps closely to how humans perceive and describe colors. Therefore, it is much easier to differentiate between objects of similar color but different brightness and saturation levels. For that, people would like to choose HSV space as one pre-trained part to provide the network with more information about the color composition of the input images, which can help it to recognize objects more accurately.

The HSV module gains more information from an image in pre-training. The first step is to modify the preprocessing step to include the HSV module. This module can be used to convert the RGB input images into the HSV color space. This is designed to abstract the color features of the input images more effectively than the RGB space. Then normalization and data augmentation are applied for the HSV images. Finally, all the datasets those in the target color space area are filled to make color discrimination. So the HSV method has been completed to extract color features. In conclusion, the HSV model is more sensitive to color differences when extracting features in shallow layers, and particularly focusing on the white area and its periphery. After that, the picture is converted to a grayscale, and the color falling in the area is filled to enhance the sensitivity of the model to the fixed area. The conversion formulas are as follows:

$$
H = \begin{cases} 0° \\ 60° \times \left(\frac{G'-B'}{\Delta} \, mod \, 6 \right), C_{max} = R' \\ 60° \times \left(\frac{B'-R'}{\Delta} + 2 \right), C_{max} = G' \\ 60° \times \left(\frac{R'-G'}{\Delta} + 4 \right), C_{max} = B' \end{cases} \tag{1}
$$

$$
H = \begin{cases} 0, C_{max} = 0 \\ \frac{\Delta}{C_{max}}, C_{max} \neq 0 \end{cases} \tag{2}
$$

$$
V = C_{max} \tag{3}
$$

where, H is the polar angle of polar coordinates, S is the polar axis length, and V is the height of the axis in the cylinder. $\Delta = Cmax - Cmin$ is the distance difference between the maximum and minimum value of a color, which consists of the original coordinate. While HSV only can be used the range is [0, 1], so (R, G, B) transforms to (R', G', B') firstly.

2.2 Contextual Attention Mechanism

After the CSPDarknet53 feature extraction, the Convolutional Block Attention Module (CBAM) is added to capture the contextual information of the small object. As shown in Fig. 2, the module can divide into two types of attention: the channel attention module (CAM) and the spatial attention module (SAM). The CAM performs channel-wise feature recalibration to adaptively rescale feature maps based on their importance, whereas the SAM performs spatial-wise feature recalibration by selectively attending to informative spatial regions.

In CAM, global maximum pooling and global average pooling should be built based on width and height at first. Then, with a 1×1 convolution reducing dimension and raising the same dimension, a multi-layer perceptron (MLP) network is constructed. After pooling and MLP, the output features are added element-wise and then activated by sigmoid to generate the final channel attention feature map, which is shown as follows:

$$
M_c(F) = \sigma(MLP(AvgPool(F))) + MLP(MLP(MaxPool(F))) \tag{4}
$$

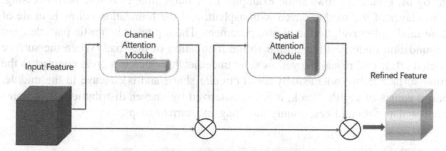

Fig. 2. The overview of CBAM. This module includes two sub-modules: channel attention module and spatial attention module. The final refined feature depends on the two sub-modules weighted results.

$$= \sigma(W_1(W_0(F_{avg}^c)) + W_1(W_0(F_{max}^c)) \tag{5}$$

where, F is original feature map, $AvgPool(*)$ is the global average pooling function, $MaxPool(*)$ is the maximum pooling function of the office, W is target weight, and σ is sigmoid function.

Different with CAM, maximum pooling and average pooling are performed on input feature graphs in the channel dimension of SAM, and then the two pooled feature graphs are stacked in the channel dimension. Then, the shape of the feature graph changes from [b, 2, h, w] to [b, 1, h, w] by using 7×7 convolution kernel fusion channel information. Finally, the convolution result is normalized by the sigmoid function to the spatial weight of the feature graph, and then the input feature graph and the weight are multiplied. The formula is as follows:

$$M_s(F) = \sigma(f^{7\times7}([AvgPool(F); MaxPool(F)])) \tag{6}$$

$$= \sigma(f^{7\times7}([F_{avg}^s; F_{max}^s])) \tag{7}$$

In short, with the contextual attention mechanism, the model can abstract and fuse more small objects' features from the contextual features and information to improve the accuracy of object detection.

3 Experiment

3.1 Datasets

The plastic runway surface defects is similar to road damage, due to both explored in the air for a long time. And there is not enough data of plastic runway surface for training the model. The published datasets, ATULYA KUMAR and DATACLUSTER LABS, are used. But these datasets only have labeled with single damage. That means it is necessary and initial to filter out carefully and label them again to classify a crack of a hole with labelImg. And the plastic data set is collected from our school playground, and labeled them by us. Figure 3 shows some examples of damage images before preprocessing. The top layer of the road is paved with asphalt, and the foundation below is made of yellow sand and gravel. So the plastic pavement. The top layer is plastic particles, but the foundation underneath is similar to the foundation of the road. When the surface is peeled off, it will reveal the yellow sand underneath, which is a great contrast to the normal surface. Also, hole usually has a circular shape and is concave in the middle. When it comes to say the crack, it is characterized by uneven distribution. It contains vertical and horizontal lines, mainly into long and narrow strips.

(a) damage on road   (b) damage on road

(c) damage on plastic runway surface   (d) damage on plastic runway surface

Fig. 3. Some examples of damage images before preprocessing. The two graphs, (a) and (b), are derived from ATULYA KUMAR, one published dataset road damage defects. The two pictures, (c) and (d), are from the playground taken independently by the school.

The trained model is expected to resolve the problem of detecting small-scale objects in the complex background. It is means that Mosaic algorithm is applied in pretreatment process, after the module of HSV. There are several advantages to use the Mosaic algorithm. It can reduce irrelevant target in the picture and satisfied the need for accuracy with small targets. At the same time, Mosaic algorithm makes the model more robust to complex background while increasing the complexity of the background. The module of HSB and enhancement effect of Mosaic algorithm is shown in Fig. 4. The input image is first extracted from the pixel values in each channel and then saved in RGB space. After that, the RGB space is converted to the corresponding HSV space, and the brightness attribute is added to the part that needs attention. All images will be preprocessed by the above HSV module. Then use the Mosaic algorithm to enhance data. Four images are randomly clipped and scaled, and then spliced together. Finally, there are 300 images in our data set. The ratio of the public set to the self-capturing set is 10:1. It contains 3 categories: crack, hole and cracks and holes. Each category in the dataset is splitted into 90% training data and 10% testing data. The size of the supplied image is fixed to 640 × 640 size by size scaling. Experiments are conducted on a rental PC container (https://www.autodl.com/) with a 2.4 GHz CPU and 50 G RAM with Ubuntu 18.04LTS operating system. The environment uses the PyTorch 1.8.1 deep learning framework and is accelerated in CUDA 11.1 in NVIDIA GeForce RTX 3090 GPU devices with 80GB of GPU memory.

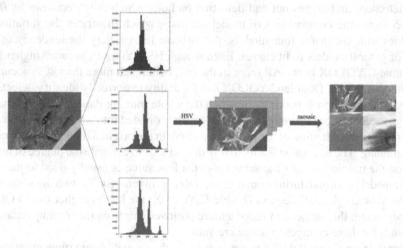

Fig. 4. The color histogram distribution of a single image is obtained, and then the brightness pixels are added to the areas that need to be focused. Also, the mosaic method is applied to collage combining multiple pictures.

3.2 Results and Analysis

All models are trained on damage runway data set and our school playground one. Also, tested on the combined data set. After training, Pascal VOC is used to test all of the models and obtain each module's detection results. Table 1 describes the performance of each model. The numbers in bold are the highest values. AP, Recall, F1-score, and mAP, are chose for comparing the performance of machine learning models. The threshold value of IOU is 0.5. AP (Average Precision) and mAP (mean Average Precision) measure the accuracy of the model in predicting the correct category for each object in an image.

As shown in Table 1, our model Double-C YOLOX has a higher mAP score and precision than the other models, which indicates that the proposed model is more accurate at detecting objects. Our model outperforms YOLOX by 2.82%. For crack, the F1 of our model increases the percentage of 1, with the score 0.794 and 0.784. For hole, our

Table 1. Ablation experiments.

Model	YOLOX	HSV	CBAM	F1(crack)	F1(hole)	mAP
1	✓			0.7848	0.8824	0.8017
2	✓	✓		0.7536	**0.9333**	0.8216
3	✓		✓	0.7368	0.8800	0.7790
4	✓	✓	✓	**0.7941**	0.9216	**0.8229**

model is slightly lower than YOLOX + HSV. It indicates that CBAM may decrease the performance of hole detection. But CBAM significantly improves the performance of crack detection, and brings not bad detection on hole, which only decreases by 0.01. Figure 5 shows the comparisons of model accuracy, which illustrates the variation of AP values with epoch for four models. All of these lines satisfy the tendency of AP values of general models to fit curves. Even though HSV YOLOX is much higher than the Double-C YOLOX in the AP score in the end, both reach more than 80% accuracy. The latter proves that Double-C YOLOX has the ability to correctly identify defects. In deep learning, one epoch is one cycle when the whole training dataset is used. Thanks to a large amount of pre-training images, an epoch is divided into several small batches. Figure 6 shows the situation of the Model loss curve in the original model and our model during training. The loss value is a relatively direct index to evaluate the degree of model fitting on the training set. At the same time, the loss value is usually used to measure whether model is normal during convergence training. In this task, the two lines converge and tend to be stable. Although in Double-C YOLOX, the loss is higher than YOLOX at the beginning, the former can judge a more positive sample on the training dataset. It also stands for these changes in usage are valid.

After training the model, this paper uses the highway and plastic runway images in the actual scene to test. The detection results of highway and plastic surface are shown. The improved model: Double-C YOLOX; not only can detect the damage of crack on the highway, but also detects the hole target with a very small area. But the original YOLOX failed to do this task.

The comparison of YOLOX and Double-C Y'OLOX in detecting cracks is displayed in two rows. In Fig. 7, there are two cracks, and YOLOX only detects the middle one, while Double-C YOLOX accurately detects both cracks. The second row illustrates high accuracy of Double-C YOLOX.

In Figs. 8 and 9, these two sets of images represent defects of different sizes. Both can locate hole and classify in correct group, but Double-C YOLOX shows higher confident score. Although both models detect the same number and location of damages, Double-C YOLOX exhibits higher confidence scores of 77.8% and 83.1%, compared to YOLOX's 58.8% and 68.9% for the same holes. The results demonstrate that Double-C YOLOX is suitable for detecting different sizes of damage as well.

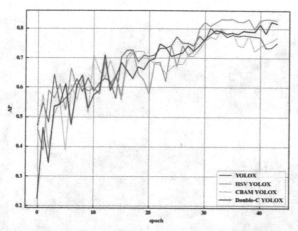

Fig. 5. Comparisons of four models' accuracy. In each model, the x-coordinate data is composed of that the first 280 generations output AP values every 10 generations. The AP value was output once from generation 281 to 285. The last 286 to 300 generations are trained, and each generation outputs an AP value.

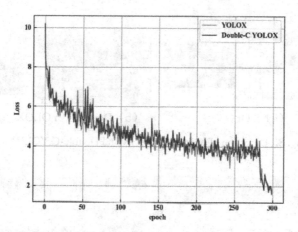

Fig. 6. Loss curve of YOLOX and Double-C YOLOX.

At the same time, more cracks are detected by our model in Fig. 10. Especially for the plastic track, our model correctly identifies the hole at the bottom of the picture, but YOLOX incorrectly classified it as a crack. Additionally, Double-C YOLOX can detect the small crack which is missed by YOLOX. More important, the former can detect more cracks at the bottom of the image.

Therefore, Double-C YOLOX model can identify damages and classify correctly cracks and holes, which caters to the need for localization and classification tasks defined by target detection. In addition, compared with YOLOX, the improved model can identify more small targets to show the advantages of the improved model.

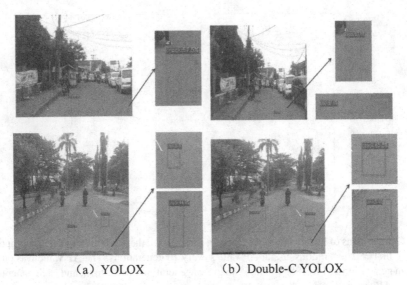

(a) YOLOX   (b) Double-C YOLOX

Fig. 7. Detection of cracks by YOLOX and Double-C YOLOX.

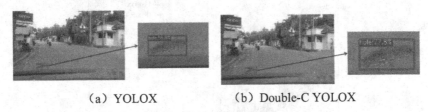

(a) YOLOX   (b) Double-C YOLOX

Fig. 8. Detection of holes by YOLOX and Double-C YOLOX.

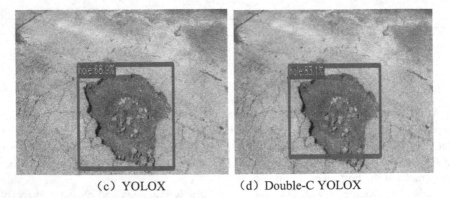

(c) YOLOX   (d) Double-C YOLOX

Fig. 9. Detection of holes by YOLOX and Double-C YOLOX.

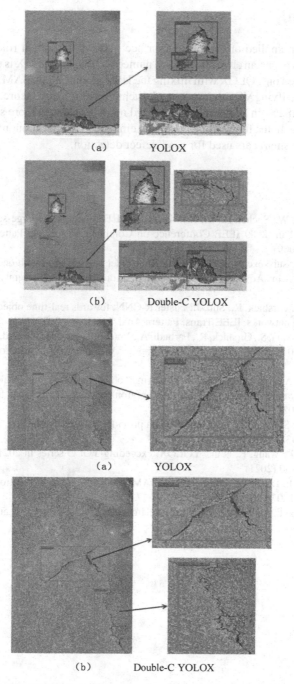

(a)   YOLOX

(b)   Double-C YOLOX

(a)   YOLOX

(b)   Double-C YOLOX

Fig. 10. Detection testing of cracks and holes on plastic surfaces.

4 Conclusion

Damage detection applied on the plastic surface plays an important role in creating a digital city. In this paper, an effective method named Double-C YOLOX is proposed. This model is built based on YOLOX with mixing the HSV module and CBAM. After training 300 epochs and verifying VOC data, our model achieves a high mAP score, improving the original one by 2.82%. In addition, the proposed model can detect more small targets of the plastic surface. In the future, this model can embedded on the autonomous machines, such as drones or smart cars used for road defect detection.

References

1. Deng, J., Dong, W., Socher, R., Li, L.J., Li, K., Li, F.F.: ImageNet: a large-scale hierarchical image database. In: 2009 IEEE Conference on Computer Vision and Pattern Recognition, pp. 248–255 (2009)
2. Krizhevsky, A., Sutskever, I., Hinton, G.E.: ImageNet classification with deep convolutional neural networks. In: Advances in Neural Information Processing Systems, pp. 1097–1105 (2012)
3. Ren, S., He, K., Girshick, R., Sun, J.: Faster R-CNN: towards real-time object detection with region proposal networks. IEEE Trans. Pattern Anal. Mach. Intell. **39**, 1137–1149 (2017)
4. Redmon, J., Divvala, S., Girshick, R., Farhadi, A.: You only look once: unified, real-time object detection. In: 2016 IEEE Conference on Computer Vision and Pattern Recognition (CVPR) (2016)
5. Chen, G., et al.: A survey of the four pillars for small object detection: multiscale representation, contextual information, super-resolution, and region proposal. IEEE Trans Syst Man cybern **52**, 936–953 (2020)
6. Chen, Y., et al.: Stitcher: feedback-driven data provider for object detection, 2. arXiv preprint arXiv:2004.12432 (2020)
7. Ge, Z., Liu, S.T., Wang, F., et al.: YOLOX: exceeding YOLO series in 2021. arXiv preprint arXiv:2107.08430 (2021)
8. Woo, S., Park, J., Lee, J.-Y., kweon, I.S.: CBAM: convolution block attention module. arXiv preprint arXiv:1807.06521 (2018)
9. Joseph, R., Farhadi, A.: Yolov3: an incremental improvement. arXiv preprint arXiv:1804.02767 (2018)

Improved War Strategy Optimization Algorithm Based on Hybrid Strategy

Jiacheng Li[1]([✉]), Masato Noto[1], and Yang Zhang[2]

[1] Department of Electrical, Electronics and Information Engineering,
Kanagawa University, Yokohama, Japan
{lijiacheng,noto}@kanagawa-u.ac.jp
[2] Graduate School of Science and Engineering, Hosei University, Tokyo, Japan
yang.zhang.6j@stu.hosei.ac.jp

Abstract. The Standard WSO algorithm has several shortcomings, including uneven distribution of initial population, slow convergence speed, and weak global search ability. To address these issues, the present study proposes an improved War Strategy Optimization (WSO) based on hybrid strategy. To begin with, the initialization of the population was done using hypercube sampling. Additionally, diversification of the population during iteration process was achieved by adopting sine/cosine strategy, Cauthy mutation and backward learning strategy. Furthermore, to enhance capabilities in global search and local development, operator retention strategy from simulated annealing algorithm was employed. Finally, three test function optimization experiments were conducted which demonstrated that the proposed war strategy optimization algorithm based on hybrid strategy significantly improves both optimization accuracy and convergence speed.

Keywords: WSO Algorithm · Hypercube Sampling · Sine/Cosine Strategy · Cauthy Mutation · Backward Learning Strategy

1 Introduction

Intelligent optimization algorithms encompass various techniques such as natural bionic optimization, evolutionary algorithm, plant growth simulation algorithm, and swarm intelligent algorithm. Researchers are continuously developing new algorithms that can be applied to diverse fields including mechanical control, image processing, and path planning [3]. This study specifically focuses on War Strategy Optimization (WSO), a novel intelligent optimization algorithm proposed in the last two years. WSO outperforms traditional optimization algorithms like Genetic Algorithm (GA) [5], Particle Swarm Optimization (PSO) [10], White Shark Optimizer (WSO) [8], and other swarm intelligent algorithms.

WSO was proposed by Ayyarao et al. [2] in 2022, and it is a new metaheuristic optimization algorithm based on ancient war strategies. In the same year, Ayyarao et al. [1] explained that ESO simulated two popular war

J. Li et al. (Eds.): 6GN 2023, LNICST 554, pp. 149–159, 2024.
https://doi.org/10.1007/978-3-031-53404-1_13

strategies: attack and defense. In order to improve the convergence and robustness of the algorithm, they designed another new weight renewal mechanism and
weak army policy, which estimated parameters of the solar photovoltaic model
to prove superiority. In addition, XU Jianwei et al. [9] proposed a war strategy optimization algorithm and chameleon swarm algorithm (CSA). In order to
raise time-series prediction accuracy of sediment runoff, they construct a time-
series forecast model and a comparative analysis model for multi-algorithms.
The results showed that CSA has excellent optimization accuracy and global
search ability.

Considering that few scholars discuss war strategy algorithm, this study aims
to improve the algorithm by using a hybrid strategy. To be specific, QIU Xingguo
et al. [6] presented the white shark optimizer through applying low-deviation
SOBOL sequence, new weight improvement formula, and local Cauchy mutation. However, their results showed that the convergence speed and optimization accuracy of the algorithm need further improvement. HE Qing et al. [4]
designed a grasshopper optimization algorithm (GOA) that combines Cauchy
mutation and uniform distribution. Their findings showed an increase in both
convergence accuracy and speed of the algorithm. In addition, WANG Weijun
et al. [7] improved upon the intelligent water drop algorithm by slowing down
its convergence speed through adding theory of second best set and contraction
factor which boosted its effectiveness significantly.

In summary, the WSO algorithm has not been extensively analyzed in previous research. Although new algorithms have been developed, they still have some
shortcomings that require further research and improvements. Against this back-
drop, this study proposes an improved hybrid war strategy optimization algorithm to enhance the optimization ability of the algorithm and prevent it from
getting stuck in local optima. The effectiveness and superiority of the proposed
algorithm are verified through experiments.

2   Algorithm Description

2.1   War Strategy Algorithm

The War Strategy Algorithm draws inspiration from the strategic operations of
armies during ancient wars. It models war strategy as an optimization process,
where each soldier dynamically moves towards the best value. The algorithm
incorporates two popular war strategies-attack and defense-while also introducing a new weight renewal mechanism and a replacement strategy for weak soldiers (those with low fitness). These additions aim to enhance convergence and
robustness. In summary, the main strategy of WSO is as follows:

(1) Attack

Two war strategies were simulated. In the first case, soldiers updated their positions based on the locations of the King and commander.

$$X_i(t+1) = X_i(t) + 2\rho(C - K) + \rho\left(W_i K - X_i(t)\right) \tag{1}$$

where $X_i(t+1)$ is the new position of the soldier in the $t+1$ iteration; $X_i(t)$ means the new position of the soldier in the t iteration; C represents the position of the commander; K is the position of the King; ρ is a random number between 0 and 1; W_i is the weight of the King's position.

If $W_i > 1$, then $W_i K - X_i(t)$ is outside the current position of the King. Therefore, updated position of soldiers is outside the commander's position. If $W_i < 1$, then $W_i K - X_i(t)$ is between the King and the current soldier. The soldier's updated position is closer to his original position than before.

Figure 1 illustrates the update mechanism for the attack model. The King, who is in a favorable position to launch large-scale attacks on the enemy, is represented by the soldier with the greatest attack power or strongest physical fitness. At the onset of war, all soldiers have equal military ranks and weights. As soldiers successfully execute strategies, their military ranks will rise. However, as the war progresses, all soldiers' ranks and weights will be updated based on their success in implementing strategies. By the end of the war, positions of both commanders and soldiers are still close to achieving victory.

(2) Rank and weight update strategy

Soldier rank R_i rests with his attack force (fitness value) on the battlefield.

The position update of each search agent depends on the interaction between the King's position and the ranks of commanders and soldiers. The rank of each soldier is determined by their contributions on the battlefield, which in turn affects their weight coefficient W_i.

Provided that the attack force F_n (fitness) of a soldier's new position is lower than the attack force F_p of their current position, the soldier will maintain the current position.

$$X_i(t+1) = (X_i(t+1)) \times (F_n \geq F_p) + (X_i(t)) \times (F_n < F_p) \qquad (2)$$

If the soldier updates his position successfully, the rank will be upgraded, expressed by:

$$R_i = (R_i + 1) \times (F_n \geq F_p) + (R_i) \times (F_n < F_p) \qquad (3)$$

The weights of soldiers change based on their attack force (fitness value), and this weight update is described mathematically as:

$$W_i = W_i \times \left(1 - \frac{R_i}{Max\_iter}\right)^{\alpha} \qquad (4)$$

In the formula, F_n is the attack force (fitness value) of a soldier's new position; F_p means the attack force (fitness value) of a soldier's previous position; R_i refers to the rank of the i-th soldier; α is an exponential change factor.

(3) Defense strategy

The update of a soldier's position depends on the positions of the King, commander, and other soldiers nearby. However, their military rank and weight remain unchanged. Soldiers will adjust their positions based on those around them and prioritize protecting the King without compromising their ability to win battles.

$$X_i(t+1) = X_i(t) + 2\rho\,(K - X_{rand}(t)) + \rho \times W_i \times (C - X_i(t)) \qquad (5)$$

where $X_{rand}(t)$ is the random position of a soldier at t-time iteration, and meanings of other parameters are unchanged. This new war strategy involves positioning random soldiers and explores a greater number of search spaces compared to previous strategies. For large W_i values, soldiers will take great steps and update their positions. Otherwise, smaller steps will be adopted.

(4) Replace or resettle weak soldiers

During each iteration, soldiers with the lowest attack force (fitness) will be replaced. It is unrealistic to expect no casualties in a war, and weak soldiers should be easily replaced by selecting a random soldier.

$$X_w(t+1) = Lb + \rho \times (Ub - Lb) \qquad (6)$$

The second method to improve the convergence performance of the algorithm is by placing weaker soldiers in the middle of the entire army on the battlefield, thus protecting them.

$$X_w(t+1) = -(1 - \rho') \times (X_w(t) - median(X)) + K \qquad (7)$$

$X_w(t+1)$ is the position of a weak soldier replaced or resettled by $t+1$ times; Ub and Lb are upper and lower limit values of search space; $X_w(t)$ represents the position of a weak soldier at t-time iteration; ρ' means a random number evenly distributed between 0 and 1; $median(\ )$ refers to the median function.

2.2   Improved War Strategy Algorithm

Hypercube Sampling. Placing soldiers according to war strategy can lead to a centralized arrangement due to uncertain enemy positions, increasing the risk of falling into local extremum. To avoid this, hypercube sampling is recommended as it is a layered Monte Carlo method that allows for uniform sampling of multi-dimensional space with only a small number of samples.

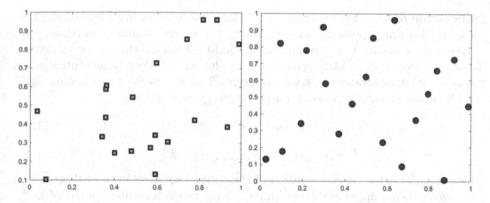

Fig. 1. Random sampling distribution.   **Fig. 2.** Hypercube sampling distribution.

The figures above show that hypercube sampling results in a more even distribution of data points and a diverse population. This suggests that good initial solutions can be obtained.

Sine/Cosine Search Strategy. In the algorithm for war strategy, the update of soldier positions is primarily influenced by attack force, resulting in a rough optimization effect. To enhance convergence accuracy and optimization effect, we introduce the sine/cosine search strategy. After soldiers change and update their positions based on those of nearby soldiers and the King, a sine/cosine search is performed to obtain an optimal feasible solution. The mathematical expression of sine/cosine search strategy is as follows (Fig. 2):

$$X_i(t+1) = \begin{cases} X_i(t) + r_1 \times \sin r_2 \times D & r_3 < 0.5 \\ X_i(t) + r_1 \times \sin r_2 \times D & r_3 \geq 0.5 \end{cases} \tag{8}$$

$$D = \{|r_4 \times X_i(t) - X_i(t)| \quad (i = 1, 2, \ldots d) \tag{9}$$

$$r_1 = a - a \times \frac{t}{M} \tag{10}$$

$$r_2 \in (0°, 360°) \tag{11}$$

where r_1 increases as the number of iterations M increases; a is a constant with a value of 2 in the experiment; r_3 and r_4 are random numbers evenly distributed on $[0, 1]$. The sine/cosine search aims to effectively prevent algorithm from being early-maturing, enhance convergence accuracy to some extent, and improve efficiency of each iteration.

Integration of Cauthy Mutation and Backward Learning Strategy. The backward learning approach involves finding the reverse solution based on the current one, evaluating and comparing it, and then saving the better solution. To assist individuals in identifying optimal solutions, this study incorporates a backward learning strategy into war strategies. This is expressed mathematically as: The mathematical expression for this strategy is as follows:

$$X'_{best}(t) = ub + r \oplus (lb - X_{best}(t)) \tag{12}$$

$$X_{i,j}^{t+1} = X'_{best}(t) + b_1 \oplus (X_{best}(t) - X'_{best}(t)) \tag{13}$$

where $X'_{best}(t)$ is the reverse solution of the optimal solution in the t-th generation; ub and lb are upper and lower limits; r is the random number matrix of $1 * d$ (d is space dimension) that is evenly distributed by obeying $(0, 1)$ standard; b_1 represents the control parameter for information exchange, expressed as follows:

$$b_1 = (iter_{\max} - t/iter_{\max})^t \tag{14}$$

Cauthy mutation originates from Cauthy distribution, and probability density of one-dimensional Cauthy distribution is as follows:

$$f(x) = \frac{1}{\pi} \cdot \frac{a}{a + x^2}, x \in (-\infty, +\infty) \tag{15}$$

When the value of "a" is equal to 1, the distribution is referred to as a standard Cauchy distribution.

The purpose of introducing the Cauchy distribution is to update target position methods, utilize the disturbance ability of the Cauchy operator, and enhance the global optimization performance of the algorithm.

$$X_{i,j}^{t+1} = X_{best}(t) + cauchy(0, 1) \oplus X_{best}(t) \tag{16}$$

where $cauchy(0, 1)$ is a standard Cauchy distribution. The random variable generating function of Cauchy distribution is $\eta = \tan[(\xi - 0.5)\pi]$.

The Cauchy distribution can be extended at both ends to generate random numbers that are far from the original point. This property can significantly disrupt soldiers, but equipping them with the ability to quickly avoid local traps can mitigate this issue. Additionally, compared to other functions, the Cauchy distribution has a lower highest peak and can reduce the time individuals spend searching for spaces, thereby accelerating algorithm convergence.

Preservation of Excellent Individuals. In order to enhance diversity, the simulated annealing algorithm incorporates a probabilistic leap. This allows for not only the retention of excellent individuals but also the preservation of non-excellent individuals with a certain probability, thereby preventing it from getting stuck in local optima.

Simulated annealing is an algorithm that allows for fault tolerance by accepting suboptimal solutions with a certain probability, known as the acceptance probability of the new solution. This probability is influenced by both the current temperature and the fitness difference between old and new solutions. As a general rule, lower temperatures result in lower acceptance probabilities, while greater differences also lead to lower acceptance probabilities. This relationship can be expressed mathematically using formula (17):

$$P = \begin{cases} 1 & E\left(x_{new}\right) > E\left(x_{old}\right) \\ \exp\left(-\frac{E(x_{new})-E(x_{old})}{T}\right) & E\left(x_{new}\right) \leq E\left(x_{old}\right) \end{cases} \tag{17}$$

where $E\left(x_{new}\right)$ and $E\left(x_{old}\right)$ represent fitness of new and old solutions respectively;

T means annealing temperature;

P is accepting probability.

The selection mode for acceptance probability follows the Metropolis criteria.

Once the acceptance probability is determined, random numbers between 0 and 1 are generated. If the random number is greater than the acceptance probability (P), no new solution will be accepted. However, if it is less than P, a new solution will be accepted.

The following are the steps for optimizing a war strategy using an improved algorithm:

(1) Initialize parameter of population, such as the number of populations n, maximum iteration times, etc.

(2) Calculate the fitness value of each soldier, and find out the current optimal fitness value and the worst fitness value of individuals, and their corresponding position.

(3) Evolve soldiers in accordance with war strategy and sine/cosine strategy.

(4) Generate a random number r_1. If r_1 is greater than 0.15, apply Cauthy mutation to soldiers. If the value is less than 0.5, perform backward learning strategy to calculate fitness values and retain positions of current best individuals along with other individuals.

(5) Update positions of soldiers, calculate attack force of each soldier, and rank the fitness value of each soldier.

(6) Simulation annealing helps retain outstanding individuals and output optimal solutions.

(7) Determine whether it reaches the number of iterations. If so, proceed to next step, or jump to Step (2).

(8) The program ends and optimal results are output.

2.3   Comparison of Algorithm Verification Function

In order to demonstrate the performance advantages and disadvantages of the proposed algorithm (MWSO), three test functions were utilized for optimization verification. The WSO (war strategy algorithm), GWO (gray wolf optimization algorithm), PSO (particle swarm algorithm), and ABC (artificial bee colony algorithm) were also included in Table 1 for comparison purposes. Each algorithm was run independently ten times, with a population size of 200 and maximum iteration number of 200. Algorithm performance was evaluated using indicators such as optimal simulation results, represented by minimum values and standard deviations, which are presented in Table 2.

Table 1. Benchmark function test table.

Function		Dimensionality	Variable range	Optimal value
F1	$f_1(x) = \sum\limits_{i=1}^{n} \left([x_i + 0.5]^2\right) + \sum\limits_{j=1}^{i} |\sin x_i| + \prod_{i=1}^{n} |x_i|$	30	$[-100, 100]$	7.5
F2	$f_2(x) = \sum\limits_{i=1}^{n} \left(\sum\limits_{j=1}^{i} x_i\right)^2 * \sum\limits_{j=1}^{i} |x_i| + \sum\limits_{j=1}^{i} |\sin x_i|$	30	$[-100, 100]$	0
F3	$f_3(x) = \sum\limits_{i=1}^{n} \left(\sum\limits_{j=1}^{i} x_i\right)^2$	30	$[-100, 100]$	0

Table 2 presents the average optimal value as an evaluation indicator for 13 test functions. The results show that the WSO algorithm outperforms other algorithms in optimizing F1, F2, and F3. Additionally, this algorithm demonstrates a faster convergence speed and better ability to escape local optima compared to others proposed in the study. Figure 3 displays the convergence curves of each algorithm, indicating that MWSO provides the fastest optimization process for three benchmark functions and effectively reduces optimization time.

Table 2. Comparison with other algorithm performance

		MWSO	WSO	GWO	PSO	ABC
F1	mean	7.50E+00	7.50E+00	7.50E+00	5.35E+01	1.33E+17
	std	1.29E−15	1.26E−15	2.22E−14	2.04E+03	2.60E+20
F2	mean	4.67E−295	9.29E+02	9.64E−15	1.03E+02	1.08E+04
	std	0.00E+00	8.68E+03	3.72E−11	1.29E+02	6.00E+03
F3	mean	3.82E−188	2.52E+04	2.86E−05	2.18E+02	2.92E+04
	std	0.00E+00	1.06E+04	1.43E−03	3.25E+02	3.64E+03

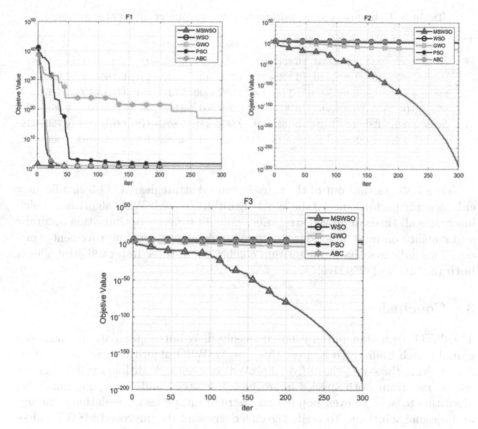

Fig. 3. Iteration curve.

2.4 Analysis of Impact of Different Improvement Strategies on Algorithm Performance

This study proposes three improvement strategies for enhancing the performance of the MSWSO algorithm. These strategies include WSO-1, which uses hypercube sampling initialization population; WSO-2, which employs a sine/cosine search strategy; and WSO-ch13, which combines Cauchy variation with backward learning strategy. To evaluate the impact of these strategies on optimization performance, we used the same parameter settings as above to optimize and solve three benchmark functions listed in Table 1. The optimization effect of these improved strategies is presented in Table 3.

Table 3. Comparison of algorithm performance under different strategies

		*	WSO	WSO-1	WSO-2	WSO-3	WSO-4	WSO-5
F1	mean	7.50E+00	2.22E+03	6.40E+02	7.70E+02	4.43E+02	6.38E+02	4.71E+02
	std	1.87E−15	1.28E−10	3.12E−14	1.91E−13	1.31E−13	1.79E−13	1.69E−13
F2	mean	4.77E−192	5.10E−100	1.72E−190	2.36E−190	5.51E−191	4.08E−190	3.21E−190
	std	0.00E+00	0.00E+00	0.00E+00	0.00E+00	0.00E+00	0.00E+00	0.00E+00
F3	mean	1.09E−120	3.70E−109	3.19E−119	2.98E−119	5.80E−119	1.05E−118	6.73E−119
	std	5.05E−107	3.00E−105	3.74E−105	1.14E−105	4.58E−105	2.14E−105	4.40E−105

Table 3 shows that out of the three improved strategies, WSO-3 significantly enhances the performance of the WSO algorithm. The MSWSO algorithm, which integrates all three improved strategies, exhibits superior optimization accuracy and stability on most functions compared to using only one improvement strategy. This indicates that an algorithm combining all three improved strategies is both rational and effective.

3   Conclusion

The WSO algorithm initially has unevenly distributed populations and weak global search ability. To address this, the MSWSO algorithm was designed by synthesizing the strengths of five improved strategies. Additionally, it integrates advantages from both species improvement strategy and simulation annealing algorithm to achieve even population distribution, precise calculation accuracy, and optimal solutions. To verify the effectiveness of the improved MSWSO algorithm, three benchmark functions were used for testing and verification against other algorithms or different strategies within the same algorithm. The results showed that MSWSO had a faster convergence speed with significantly higher optimization accuracy in most test functions compared to WSO based on a single improvement strategy. This indicates that integrating five improved strategies is both rational and effective for optimizing solutions.

References

1. Ayyarao, T.S., Kumar, P.P.: Parameter estimation of solar PV models with a new proposed war strategy optimization algorithm. Int. J. Energy Res. **46**(6), 7215–7238 (2022)
2. Ayyarao, T.S., et al.: War strategy optimization algorithm: a new effective meta-heuristic algorithm for global optimization. IEEE Access **10**, 25073–25105 (2022)
3. Gao, Y., Yang, Q., Wang, X., Li, J., et al.: Overview of new swarm intelligent optimization algorithms. J. Zhengzhou Univ. (Eng. Sci.) **43**(3), 21–30 (2022)
4. He, Q., Lin, J., Xu, H.: Hybrid Cauchy mutation and evenly distributed grasshopper optimization algorithm. Control Decis. **36**(7), 1558–1568 (2021)
5. Kapilevich, V., Seno, S., Matsuda, H., Takenaka, Y.: Chromatin 3D reconstruction from chromosomal contacts using a genetic algorithm. IEEE/ACM Trans. Comput. Biol. Bioinf. **16**(5), 1620–1626 (2019)

6. Qiu, X., Wang, R., Zhang, W., et al.: An improved whale optimization algorithm base on hybrid strategy. Comput. Eng. Appl. **58**(1), 70–78 (2022)
7. Wang, W., Luo, W.: A research of cold-chain logistic distribution path optimization based on improved intelligent water drop algorithm. Ind. Eng. J. **20**(2), 38–43 (2017)
8. Xu, H., Zhang, D., Wang, Y.: Hybrid strategy to improve whale optimization algorithm. Comput. Eng. Des. **41**(12), 3397–3404 (2020)
9. Xu, J., Cui, D.: War strategy algorithm and chameleon algorithm optimize sediment runoff time-series sequence prediction of extreme learning machine. Water Power **48**(11), 36–42 (2022)
10. Zhang, Q., Wang, Y.: Multi-swarm collaborative particle swarm optimization algorithm based on comprehensive dimensional learning. Appl. Res. Comput. **39**(8), 8 (2022)

Quantitative Analysis on Coin Flipping Protocol

Xu Guo[✉][iD] and HaoJie Li

Shanghai DianJi University, Shanghai 201306, China
`guox@sdju.edu.cn`

Abstract. Coin tossing protocol allows two mutually trustful parties to generate relative fair random numbers. The performance and security of this protocol is of great importance where random numbers are used. This paper focuses on the analysis of the performance and security of the coin tossing protocol from quantitative perspective. Data for this study is collected using probabilistic model checking technique. The behavior of a single dice can be modeled as a discrete time Markov chain. The combination of two dice can be described as a Markov decision process since the cooperation of them will generate non-deterministic choices. This paper verifies the properties such as the generation of random results and the expected number of coin tossing. Experimental results show that the protocol can work well in normal situation. The expected number of coin tossing is 4 for a single dice, and 8 for the combination of two dice. As for the security, this paper assumes that one party has been seized by a malicious adversary, the behavior of the dice combination does have been greatly affected. The probability that the expected number of coin tossing within 8 steps cannot even exceed 0.11.

Keywords: probabilistic model checking · coin flipping · random number generation

1 Introduction

1.1 Background

Random numbers are widely used in many fields, not only including probabilistic simulation, numerical analysis, security protocol and cryptography but also in the fields of communication, gambling machines etc. The goal of random number generator (RNG) is to simulate a mathematical concept of mutually independent random variables which distribute over the interval [0,1] uniformly. There are many kinds of RNGs used in computers. In fact, the RNGs are probabilistic programs which implement certain algorithms [1,2]. The existence of such kind of random mechanism makes it rather difficult to debug or verify a program.

There are many examples of simple RNG algorithms that cannot be measured in performance (or even time complexity sometimes) by deterministic solutions. Formal verification is mainly suitable for the verification of random algorithms,

J. Li et al. (Eds.): 6GN 2023, LNICST 554, pp. 160–171, 2024.
https://doi.org/10.1007/978-3-031-53404-1_14

because testing cannot prove that the algorithms are right and as an alternative, stochastic simulation, which can thoroughly traverse state-space of the probabilistic models and provide numerical analysis results. As a formal verification technique, probabilistic model checking can verify systems that exhibit probabilistic and/or non-deterministic behaviors, it also can provide quantitative measures about the verified systems. The advantage of this technique is it will explore the state space of the model exhaustively but not repeatedly. The major goal of this paper is to model and verify a RNG algorithm based on probabilistic model checking.

This paper considers a coin-tossing protocol which is a RNG algorithm proposed by Knuth and Janson [3]. The purpose of this algorithm is only to use fair coins to imitate fair dice tossing. The correctness of this algorithm has been verified by Joe Hurd [4]. The key difference is that Joe uses a theorem prover HOL while this paper uses a probabilistic model checker PRISM [5] to verify the correctness of this probabilistic algorithm.

Model checking has been a long-standing formal approach in software verification. Normally, the system to be verified is always encoded in a model-checker based specification language which is easy to be explored automatically [6,7]. At the same time, system properties to be checked are described as formulas in temporal logics [8], such as PCTL, LTL etc. For a system design, model checking techniques can verify whether the properties of the system expressed in formal specification are satisfied. As a quantitative branch of model checking, probabilistic model checking can proceed numeric verification to systems, provide not only qualitative but also quantitative analysis results. The existing major techniques which can be used to quantitatively verify probabilistic systems are discrete event simulation and probabilistic model checking. Discrete event simulation is a technique that can only verify correctness partially which can provide approximative results by calculating the average amount of plenty of sample examples. Moreover, discrete event simulation will visit duplicate state space of the model during calculation. For these reasons, discrete event simulation technique is unsuitable for the verification of correctness and performance of coin tossing protocol. On the other hand, probabilistic model checking can encapsulate the behaviors of the targeted systems and model them using the corresponding mathematical tools. As a further step, probabilistic model checking will explore the state space of the model thoroughly and provide accurate numerical results instead of approximate ones. Hence, probabilistic model checking technique is ideally suitable for the security and performance verification of probabilistic system. This paper researches the application of probabilistic model checking to the verification of security and performance of coin-tossing protocol.

Related Works. Simple version of coin-flipping protocol is commonly used in cryptography whenever security with abort is adequate. In an ideal situation, a coin-flipping protocol can run $O(n)$ times and generate random numbers in the range of $[1, 6]$. There are relatively few historical studies in the area of the security of coin-flipping protocol. Publications that concentrate on round-complexity more frequently adopt a full prove approach. Moran etc. [9] researched the round complexity of the simplest version of coin-flipping protocol in an ideal situation.

They proved that with the help of a trusted third party as a supervision, the r-round two-party coin-flipping protocol will have bias with minimum probability of $1/(8r+2)$. As a further step, Moran etc. refined the protocol, presented a generalized protocol, which improved the security guarantee. The refined protocol removed the trusted third party and assumed that one party is malicious, it can interfere the random number generation. They proved that for a r-round refined protocol, with the probability $(1 - 2^{-r})/4r$, the output of the honest party will be biased. Although it is a rigorous way to verify the properties of the protocol by proving, but there has been little quantitative analysis about the configuration variations of this protocol. It is almost powerless to deal with the parameter changes of the protocol. Probabilistic model checking not only can provide numeric verification results, but also can deal with the changes of the parameter of the protocol flexibly. HA Khorasgani etc. studied the round complexity of optimally-fair coin-tossing protocol [10]. Several lines of evidence suggest that as long as the protocol is unconscious to the implementation of one-way function, the protocol with an arbitrary number of rounds can not be based on one-way functions in a black-box way. As there are many aspects, such as security, correctness etc., to be considered before the practical applications of this protocol, there has been no detailed investigation of other aspects of this protocol except round complexity. Li etc. studied the security about coin tossing protocols which is based on Quantum Bit Commitment [11]. They presented a new scheme for bit commitment, including committing and decommitting. Overall, these studies provide important insights into the study about the round complexity of this protocol. Up to now, far too little attention has been paid to the performance, security and other aspects about this protocol. It is still not known whether the performance of this protocol can have effect on the generation of random numbers. Very little is known about whether different configurations of this protocol can affect the efficiency of random number generation. The specific objective of this study is to determine quantitatively about the performance and security of this protocol under the situation of different configurations. The approach to empirical research adopted for this study is probabilistic model checking, which is a quantitative branch of model checking.

The main contributions of this paper can be summarized as follows:

1. This paper verifies the security and performance of coin tossing protocol quantitatively. The past decade has seen gigantic requirements of the application of random numbers. To guarantee the security of random number generation and to improve the performance of the corresponding protocols has become a new focus. To the best of our known, data about the security and performance of coin tossing protocol is limited. The existing body of research on the verification of this protocol mainly focuses on round complexity but not includes other aspects such as security. This paper uses probabilistic model checking technique to investigate the security and performance of this protocol under different configuration situations.

2. This paper not only considers the performance of this protocol in the normal situation, but also verifies the round complexity in the biased situation. To the best of our known, the latter has not been considered in other literatures.

3. This paper discovers the factor that affect on the performance of this protocol. In normal situation, the performance is mainly measured by round complexity, i.e., the expected number of coin tossing. The lower round complexity, the better performance the protocol obtains. In the biased situation, the round complexity will rise sharply for the reason that one party will always tell a lie with probability p for the sake of itself. If the protocol run in the biased mode, the performance will decrease sharply.

2 Preliminaries

This section begins with some brief background materials on Markov decision process and the corresponding formalisms that will be used to present model.

Definition 1. Discrete probability distributions over a finite set F are defined by a function dis: F→[0,1]. For any f∈F, there holds Σdis(f) = 1. Dis(F)can represent all the distribution sets over F.

Definition 2. A DTMC is a tuple D=(S,s_0, t, lab) where:

- S is a finite set of states with the initial state s_0
- t is a transition probability matrix, t: S×S→[0,1]
- lab: S→ 2^{AP} is a function labeling states with atomic propositions.

A discrete time Markov Chain is a state transition system augmented with probability. It is suitable to describe the behaviors of a single dice. As a common case, two dice are always used as a pair. To describe the cooperation of two dice, they can be considered as two probabilistic process operate asynchronously. At the same time, the operation of two dice will induce non-deterministic choices. The most suitable formalism for the pair is Markov decision process, which is suitable to modelling systems that exhibits probabilistic and nondeterministic behaviors. Formally,

Definition 3. A Markov Decision Process (MDP) is a tuple M= (S,s_0, Act, steps, L) where:

- S is a finite set of states with an initial state s_0,
- Act is a finite set of actions,
- steps : S → $2^{Act \times Dis(S)}$ is the transition probability function, where Dis(S) is the set of discrete probability distributions over the set S.
- L: S → 2^{AP} is a label function where AP is the set of atomic propositions.

A path is a concrete behavior of a specific system modeled by an MDP. It solves both non-deterministic and probabilistic choices of the MDP. Formally,

Definition 4. A path ω through an MDP is a non-empty set of sequence of probabilistic transitions. $\omega = s_1 \alpha_1 s_2 \alpha_2 \ldots$, where $s_i \in S$, $\alpha_i \in Steps$ such that $P(s_i, \alpha_i, s_{i+1}) > 0$ for all i ≥ 1. A finite path is denoted as P_{fin}, and $|P_{fin}|$ is used to denote the length of path P_{fin}, the last state of a finite path is denoted as last(P_{fin}).

To consider the probability of some behaviors of an MDP, the first step is to solve the non-deterministic choices in an MDP which can produce a DTMC. In this way, we can calculate the probability of a certain path. Adversary is a solution to the non-deterministic choices in an MDP [12]. Formally,

Definition 5. An adversary δ of an MDP is a function which maps every finite path $\omega = s_1\alpha_1s_2\alpha_2\ldots s_n$ to an element $\delta(\omega)$ of Steps(s_n).

For a wide range of quantitative measures of DTMC or MDP models, the models can be augmented with Costs/Rewards. They are real valued numbers allocated to states or transitions. Formally,

Definition 6. Cost/Reward. For an MDP, cost/reward is a pair (ρ, ι), where
ρ :S$\rightarrow R^+$ is a vector that stands for state reward function.
ι :S$\times S \rightarrow R^+$ is a matrix that stands for transition reward function.

Properties to be verified are always specified by probabilistic temporal logics, such as PCTL(Probabilistic Computation Tree Logic) [13]. Due to the space limitations, the details about PCTL is omitted.

3   Model

It is generally accepted that the probability of coin tossing results is one half. For this reason, coin-tossing is believed as a fair way because both parties can see the results immediately which are almost completely unbiased. Due to the fairness of coin tossing, Morgan and Naor etc. proposed that a random dice number can be determined by a fair coin [14]. For a single dice, the final number it generates can be denoted as x = (R mod 6)+1, where R is a random number in the range of [1,6]. Informally, a die based on fair coins is an iteration process. The head side stands for 0 while the tail side stands for 1. One will toss a coin three times in order to obtain three binary bits which stands for an integer ranging from 1 to 6. If one obtains triple 1 or triple 0, the tossing process will restart until one can obtain the final numbers. For example, after three times of tossing, one obtains head, tail, head, it means the final number is 2. The process of a coin tossing can be modeled as a DTMC as Fig. 1 shows. The upper branch stands for head side while the lower branch stands for the tail side. The probability of obtaining head or tail side is denoted as p or 1-p in the DTMC.

Bit commitment (BC) proposed by Chailloux [15] can be used in the field of zero-knowledge proof, coin-tossing protocol etc., is the basic protocol of cryptography. BC can guarantee the fairness of coin-tossing to a certain extent. BC can be divided into two stages: commitment stage and reveal stage. In the commitment stage, the sender Alice will send two commitment data items, named w_1 and w_2, which represent the front and back side of a coin randomly. The receiver Bob must choose one from w_1 and w_2 but he has no way to learn the meaning of them. In the reveal stage, Bob tells Alice the selection result (w_1 or w_2), and Alice reveals the only content represented by the result, that is, the front or the back. At the same time, Alice sends the validation parameters to Bob and Bob can verify the correctness.

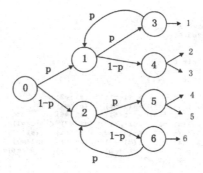

Fig. 1. Coin tossing model

In general, a coin-tossing protocol is always defined as two probabilistic polynomial-time Turing machines, denoted as (P_1, P_2). Each party can receive an input parameter 1^n. The parties will communicate in a sequence of rounds. In the end, P_1 and P_2 generates output bits r_1 and r_2 respectively. We can denote by $(r_1|r_2) \leftarrow < P_1(1^n), P_2(1^n) >$ as a process in which P_1 and P_2 communicate and generate the corresponding outputs r_1 and r_2. For an adequately large n, the pair (r_1, r_2) may be output by $< p_1(1^n), p_2(1^n) >$, it holds that $r_1 = r_2$. That means the parties will agree on common values with appropriately high probability. Beimel's proof shows that if the parties agree on a common value with probability $1/(2+\epsilon)$, then the r-round protocol has bias at least $\epsilon/(4r + 1)$ [16].

Therefore, the individual behaviors of P_1 and P_2 can be modeled as two separate DTMCs, which can generate dice number separately. Since P_1 and P_2 are concurrent threads, the operation of them will induce non-determinism. An MDP model can describe the probabilistic and non-deterministic behaviors of P_1 and P_2. The MDP model is as Fig. 2 shows.

In the MDP model, dice2 is identical to dice1, so we can in fact replace its entire definition with that of dice1 using module renaming that PRISM language supports. For the statement $s_1 = s_2$, it means to s_2 is to take the place of s_1 in the module dice1.

4   Probabilistic Model Checking

4.1   Single Dice

The first step is to verify the correctness of this protocol. For a single dice, the probability of obtaining head side or tail side is p = 0.5 and the probability of reaching the final number is $1/6$. As the experimental results are shown in table 1, it is further proved that the DTMC model can describe the behavior of a single dice. The model checker uses 4 iterations to find the reachable states. In PRISM language, this property can be described as follows:

```
mdp

const double p1=0.5;
const double p2=0.5;
module dice1
    // local state
    s1 : [0..7] init 0;
    // value of the dice
    d1 : [0..6] init 0;

    [] s1=0 -> p1 : (s1'=1) + (1-p1) : (s1'=2);
    [] s1=1 -> p1 : (s1'=3) + (1-p1) : (s1'=4);
    [] s1=2 -> p1 : (s1'=5) + (1-p1) : (s1'=6);
    [] s1=3 -> p1: (s1'=1) + (1-p1) : (s1'=7) & (d1'=1);
    [] s1=4 -> p1 : (s1'=7) & (d1'=2) + (1-p1) : (s1'=7) & (d1'=3);
    [] s1=5 -> p1: (s1'=7) & (d1'=4) + (1-p1) : (s1'=7) & (d1'=5);
    [] s1=6 -> (1-p1) : (s1'=2) + p1 : (s1'=7) & (d1'=6);
    [] s1=7 & s2=7 -> (s1'=7);
endmodule

module
    dice2 = dice1 [ s1=s2, s2=s1, d1=d2, p1=p2 ]
endmodule
```

Fig. 2. Model of two dice

$$P =?[Fs = 7\&d = k], where\ k = 1, 2, \cdots, 6. \tag{1}$$

Table 1. Verification results for a single dice.

k	probability	tossing number(ave.)
1	0.167 (1/6)	3.667 (11/3)
2	0.167 (1/6)	3.667 (11/3)
3	0.167 (1/6)	3.667 (11/3)
4	0.167 (1/6)	3.667 (11/3)
5	0.167 (1/6)	3.667 (11/3)
6	0.167 (1/6)	3.667 (11/3)

The expected number of coin tossing is also a concern. We can augment 1 to each transition in the DTMC model during the process of number generation. On obtaining the final result, the total number of transitions is the number of coin tossing. This can be calculated by reward structure. Since the final number a dice is a uniform distributed random variable on $[1, 6]$, the expected coin tossing number is a definite value. The experimental results show the number is 4, as listed in Table 1. This property can be calculated using reward/cost through the formula:

$$R =?[Fs = 7] \tag{2}$$

The expected number of coin flips is also a concern of this protocol. The number of coin flips can be represented by reward structure as follows:

$$R =?[Fs_1 = 7\&s_2 = 7] \tag{3}$$

The experimental results show that average number of coins tossing for two dice is 7.33, i.e., 22/3.

4.2   Cooperation of Two Dice

One property about this protocol is the probability that the sum of the two dice equals to a certain number. Since the sum is required not only for computer games but also for many other applications involving statistical experiments. To calculate the probability, the prerequisite is the two dice respectively achieve an exact number within equal steps. The MDP model is an asynchronous model, in which two dice can operate independently. Due to the non-deterministic behaviors of the two dice, MDP model can provide the maximum and/or minimum probability of obtaining an exact summation which distribute within the range of $[2, 12]$. The purpose of the MDP model is to describe the concurrency of the two dice, but not non-determinism. This means the maximum probability and minimum probability that the sum of the two dice is a certain number should be identical.

The property that the two dice have obtained the final numbers can be denoted as $s_1 = 7$, $s_2 = 7$ in the MDP model. Two variables d_1 and d_2 stands for the exact numbers each dice has obtained, so the property that the sum of two dice equals to a certain number can be described as

$$P =?[Fs_1 = 7\&s_2 = 7\&d_1 + d_2 = k], where \ k = 2, \cdots, 12. \qquad (4)$$

Verification results are shown in Table 2.

Table 2. Verification results for the sum of two dice.

k	P_{min}	P_{max}
2	0.028	0.028
3	0.056	0.056
4	0.083	0.083
5	0.111	0.111
6	0.139	0.139
7	0.167	0.167
8	0.139	0.139
9	0.111	0.111
10	0.083	0.083
11	0.056	0.056
12	0.028	0.028

The behavior of a single dice can be denoted as a random variable which obeys uniform-distribution on $[1, 6]$. Discrete time Markov chain is competent to represent this situation. But DTMC model is lack of the ability to describe concurrent circumstance, this paper adopts MDP to model the concurrent behaviors of the two dice. But MDP model can only provide maximum and/or minimum probability for certain event due to the existence of non-deterministic choice in the model. As for this algorithm, the two dice are mutual independent, there is no non-deterministic choice in the MDP model. That means, the maximum and minimum probability are equal. The experimental results also proved this fact.

4.3 Bias Situation

For the face-to-face situation, the fairness of the protocol can be guaranteed through supervision. But it will encounter difficulties in network environment. In the process of communication, one party may be coerced and is likely to cheat. For example, one party will cheat on its coin tossing results and tell the other party the results in favor of itself. Therefore, the correctness and security of this protocol is a major concern of this paper.

The bias coin-flipping situation can be viewed as special case of the more widespread architecture of fairness in the view of security in practical application. The experiment supposes that one party has been seized by a malicious adversary which always misstates the coin tossing result with a certain probability. For simple representation, the probability is described as p in Fig. 1. To study how the seized party will affect the security of this protocol, this paper calculates the probability that the sum of two dice reaches k, where k = 2,3,...,12 in bias situation. To discover the regularity of the change of the probability, this paper studies six different situations, where p varies from 0.2 to 0.8. Obviously, the minimum probability is 0, the maximum probability is listed in Table 3.

Table 3. The maximum probability of the sum of two dice (bias situation).

k	p=0.2	p=0.3	p=0.4	p=0.6	p=0.7	p=0.8
2	0.006	0.012	0.019	0.037	0.048	0.059
3	0.011	0.023	0.038	0.075	0.096	0.119
4	0.033	0.050	0.067	0.100	0.117	0.133
5	0.040	0.063	0.088	0.132	0.148	0.159
6	0.065	0.094	0.119	0.153	0.161	0.165
7	0.167	0.167	0.167	0.167	0.167	0.167
8	0.161	0.155	0.148	0.129	0.119	0.107
9	0.156	0.144	0.129	0.092	0.071	0.048
10	0.133	0.117	0.100	0.067	0.050	0.033
11	0.127	0.103	0.079	0.035	0.019	0.008
12	0.102	0.072	0.047	0.014	0.006	0.002

Figure 3 illustrated the trends of the maximum probability. The black dashed line shows the ideal situation, since the two parties are honest, the curve is symmetrical. The red and blue curve represents the situation that p=0.8 and p=0.2 respectively, since one party is dishonest, it can greatly promote or reduce the probability. Moreover, the counter parts of the two curves are slightly asymmetric.

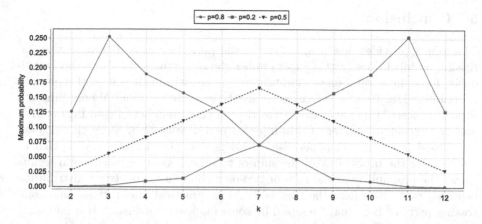

Fig. 3. The trends of the maximum probability that the sum of two dice reaches k.

In the ideal mode, the expected number of coin flips is 7.33 as the experimental results shown in section (1). Can this protocol achieve the goal that the expected number of coin flips is no more than 8 under biased situation? The property can be described as follows:

$$P_{max} =?[F \leq 8s_1 = 7\&s_2 = 7\&d_1 + d_2 = k], where\ k = 2, \cdots, 12. \quad (5)$$

Experimental results show that it is almost impossible to achieve this goal, as Fig. 4 shows, the maximum of this probability will not even exceed 0.11.

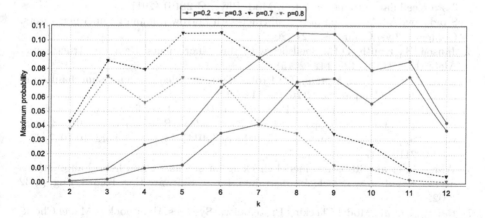

Fig. 4. The maximum probability that the expected number of coin-flips is no more than 8.

5   Conclusion

This paper provides new insights about the performance and security of coin-tossing protocol. Firstly, this paper studies the correctness of a single dice. Obviously, for a single dice, one can obtain an integer within the range of [1,6] with probability 1/6. Besides the correctness, the expected number of coin flips is a further concern, this paper calculates the expected number of coin flips is 11/3 and the expected number of coin-flips for two dice is 22/3. Since the dice are often used as a pair, this paper also considers the operation of two dice, that is, to calculate the probability of the sum of two-dice. As the experimental results show, the probability of the sum of two-dice shows a symmetrical distribution centered on 0.167 where the sum is 7. As in practical applications, one coin-tossing party or both may be seized by some malicious adversary. It is necessary not only to consider the fair situation but also to study the bias case. Under the bias situation, this paper considers one coin-toss has been seized and will tell a lie with probability p. This paper studies the maximum probability of the sum of two dice under the bias situation and compares the probability with that in the fair case. Since experimental results have shown that under the fair situation, the average number of coin-flips is 22/3, i.e., 7.33, this paper verifies the probability for the sum of two dice obtaining a certain number with 8 steps. Experimental results show that it is almost impossible for two dice to obtain certain number within 8 steps.

References

1. Bhattacharjee, K., Paul, D., Das, S.: Pseudo-random number generation using a 3-state cellular automaton. Int. J. Mod. Phys. C **28**(6) (2017)
2. Sibidanov, A.: A revision of the subtract-with-borrow random number generators. Comput. Phys. Commun., 221 (2017)
3. Janson, S., Knuth, D.E.: Shellsort with three increments. Random Structures & Algorithms **10**(1), 125–142 (2015)
4. Hurd, J., Mciver, A., Morgan, C.: Probabilistic guarded commands mechanized in HOL. Theor. Comput. Sci. **346**, 96–112 (2005)
5. Kwiatkowska, M., Norman, G., Parker, D.: PRISM 4.0: verification of probabilistic real-time systems. In: Gopalakrishnan, G., Qadeer, S. (eds.) CAV 2011. LNCS, vol. 6806, pp. 585–591. Springer, Heidelberg (2011). https://doi.org/10.1007/978-3-642-22110-1_47
6. Wang, J., et al.: Statistical model checking for stochastic and hybrid autonomous driving based on spatio-clock constraints. Int. J. Software Eng. Knowl. Eng. **4**, 32 (2022)
7. Baier, C., et al.: Model Checking Probabilistic Systems. Handbook of Model Checking (2018)
8. Rybakov, V.: Multi-modal and temporal logics with universal formula-reduction of admissibility to validity and unification. J. Logic Comput. **4**, 509–519 (2018)
9. Moran, T., Naor, M., Segev, G.: An optimally fair coin toss. J. Cryptol. **29**(3), 491–513 (2016)

10. Khorasgani, H.A., Maji, H.K., Wang, M.: Optimally-secure Coin-tossing against a Byzantine Adversary. In: IEEE International Symposium on Information Theory IEEE (2021)

11. Li, L., Shi, R.H.: A feasible and practically secure quantum bit commitment protocol. Modern Phys. Lett. A (2021)

12. Piribauer, J., Sankur, O., Baier, C.: The variance-penalized stochastic shortest path problem (2022)

13. Wang, Y., et al.: Statistical model checking for hyperproperties. In: 2021 IEEE 34th Computer Security Foundations Symposium (CSF) IEEE (2021)

14. Moran, T., Naor, M., Segev, G.: An optimally fair coin toss. J. Cryptol. **29**(3), 491–513 (2016)

15. Chailloux, A., Kerenidis, I.: Physical limitations of quantum cryptographic primitives or optimal bounds for quantum coin flipping and bit commitment. SIAM J. Comput. **46**(5), 1647–1677 (2017)

16. Beimel, A., Omri, E., Orlov, I.: Protocols for multiparty coin toss with a dishonest majority. J. Cryptol. **28**(3), 551–600 (2015)

Communications Systems and Networking & Control and Automation Systems

Research on Cost Optimization of UAV Network Routing Protocol Based on OLSR Protocol

Zhenyu Xu[✉], Xinlu Li, and Dawu Xu

Huizhou Engineering Vocational College, Huizhou, China
hitusa@126.com

Abstract. The HELLO message stands as the foundational and pivotal message structure within the OLSR protocol. It encompasses vital information, including neighbor types, primary addresses, and link state details concerning all neighboring nodes of the transmitting node. Its primary purpose lies in establishing the local link information base, creating 1-hop and 2-hop neighbor sets, and facilitating the calculation of node MPR sets during the phases of link awareness and neighbor discovery. Evidently, the ability of the HELLO message transmission frequency to keep pace with topology changes profoundly influences the protocol's communication performance. A strategic reduction in the frequency of HELLO message transmission is a means of alleviating routing control overhead. Hence, the crux of the matter becomes the adaptive adjustment of HELLO message transmission intervals in response to topology alterations. This challenge takes center stage in addressing the issue. In this study, by leveraging the comprehensive neighbor node data within the node's local link information base and considering pertinent changes in network topology—such as topology establishment, shifts in link symmetry status, and link disruptions—the methodology involves monitoring shifts in the node's local link information base. This approach, in turn, achieves an automated fine-tuning of the HELLO message transmission interval. Ultimately, the approach is subjected to testing, validation, and analysis using UAVs.

Keywords: UAV Network · OLSR Routing Protocol · Overhead Optimization

1 Background

The local link set contains comprehensive information about neighboring nodes, encompassing details such as the interface address linked to the neighbor node, the neighbor node's address, link type, expiration time, and more [1]. Changes pertaining to the network topology, whether involving topology establishment, modifications in link symmetry, or link disruptions, are all registered within the local link information base of the relevant node impacted by the topology adjustment [2]. Consequently, monitoring changes in a node's local link information base allows for the identification of topology shifts, facilitating the automatic adjustment of the interval for sending HELLO messages [3, 4].Building upon the aforementioned context, the core concept of the adaptive adjustment algorithm for the frequency of HELLO message transmission revolves around

J. Li et al. (Eds.): 6GN 2023, LNICST 554, pp. 175–187, 2024.
https://doi.org/10.1007/978-3-031-53404-1_15

maximizing the transmission interval for HELLO messages, all while ensuring that the message frequency remains responsive to timely network topology changes. This approach takes cues from variations in the node's local link information database, serving as the yardstick for determining the appropriate interval for transmitting HELLO messages [5].

The primary focus of this endeavor revolves around the dynamic shifts in the node's local link information database's membership, along with changes in the counts of three key parameters: L_SYM_time, L_ASYM_time, and L_TIME. In this context, L_SYM_time, L_ASYM_time, and L_TIME correspondingly denote the time instances linked to the states of SYM_LINK, ASYM_LINK, and LOST_LINK. These states act as identifiers for link statuses.L_SYM_time signifies that the links before this timestamp maintain symmetry, indicating bidirectional symmetric connections. SYM_LINK represents a bidirectional symmetric link, enabling mutual data packet transmission between both nodes. Conversely, L_ASYM_time designates instances when links are asymmetric, signifying unidirectional symmetric connections. ASYM_LINK signifies that a link is asymmetric, allowing message reception only from the opposite node. L_TIME points to periods when a link remains valid; LOST_LINK designates a failed link where packet exchange between the nodes on either end is impossible.The project employs a tallying approach to track changes in link status identifiers during each HELLO message transmission interval. However, the project constructs these tallying rules based on a comprehensive analysis of link alterations. Simultaneously, the project adjusts the HELLO message transmission interval proportionally. To enhance this proportional correction's accuracy, the project introduces the average change count of the local link information database within a specific unit of time [6–8]. The algorithm's fundamental process is outlined as follows:

(1) First, a link change counter (LCC) is established within each node, serving as a metric for quantifying alterations within the local link information base for every instance of HELLO message transmission.

(2) The LCC plays a pivotal role in enumerating changes during each interval when a HELLO message is sent [9]. The criteria for counting within the LCC are dissected as follows:

a) Upon the inclusion of a new element within a node's local link information base, a fresh neighbor node is introduced into the node's vicinity. However, considering the prior analysis of the OLSR protocol's link awareness phase, the creation of a symmetric link between two nodes necessitates the exchange of a minimum of three HELLO messages. To expedite the completion of the link awareness and neighbor detection phase, facilitating timely topology establishment, it becomes imperative to elevate the counting significance during this juncture. Consequently, whenever a new element integrates into the node's local link information base, the Link Change Counter (LCC) is augmented by a value of 3.

b) In situations where a link state ID shifts from states other than SYM to SYM within the local link information base, it signifies the successful establishment of a bidirectional symmetric link between the local node and the respective node. Although topology alterations are typically infrequent within brief timeframes, it remains essential to promptly notify neighboring nodes of such changes. In these scenarios, the Link Change Counter (LCC) is incremented by a value of 1.

c) When a link status ID within the local link information base transitions to ASYM_LINK, it signifies a change in the topology. This change could involve a shift from a bidirectional symmetric link between two nodes to a unidirectional one, or the reconnection of a previously lost link (LOST_LINK). It is crucial to promptly monitor such changes. In response, the sending interval of HELLO messages should be decreased to reflect the altered state. In this context, the Link Change Counter (LCC) experiences an increment of 2.

d) Whenever an operation involves the deletion of an element within the local link database, or when a link ID transitions to LOST_LINK, it signifies that a node has exited the communication range of the local node. This departure isn't expected to negatively impact subsequent link awareness and neighbor detection processes [10]. Consequently, the increment to the Link Change Counter (LCC) is set to 0.

The calculation formula for LCC within a HELLO transmission interval is expressed as follows:

$$LCC = 3 * New + 2 * ASYM + 1 * SYM \tag{1.1}$$

where, New is the number of new elements in the local link information base, and $ASYM$ is the number of times the link ID changes to $ASYM\_LINK$ The number of LINK times. The SYM is the link ID and becomes the SYM_LINK.

Where New represents the count of new elements added to the local link information base, $ASYM$ indicates the frequency of occurrences where the link ID shifts to $ASYM\_LINK$, and SYM denotes the number of instances when the link ID transitions to becoming SYM_LINK.

(3) As defined by the OLSR protocol, the HELLO message sending interval, denoted as HELLO_INTERVAL, is established at 2 s. For this project's purposes, a lower limit for the HELLO message sending interval, HI_MIN, is assumed, calculated as HELLO_INTERVAL - ΔHI. Correspondingly, an upper limit, HI_MAX, is established as HELLO_INTERVAL + ΔHI. These bounds are devised to facilitate the adaptive adjustment of the sending interval.To elaborate, within a node's unit transmission interval, the project considers the average value of the Link Change Counter (LCC), denoted as ALCC. Building on this foundation, the formula for calculating the adaptive HELLO message sending interval, HI, can be expressed as follows:

$$HI = \begin{cases} HELLO\_INTERVAL + \Delta HI, & 0 \le x < 1 \\ -2*\Delta HI * x + 2 * \Delta HI + HELLO\_INTERVAL, & 0.5 \le x < 1.5 \\ HELLO\_INTERVAL - \Delta HI, & 1.5 \le x \end{cases} \tag{1.2}$$

where $x = LCC/ALCC$.

(4) Given that the counting of the Link Change Counter (LCC) occurs within the HELLO message sending interval, a unique scenario can arise. This scenario involves the LCC value potentially becoming excessively large or too small within an HI interval. This situation could lead to the calculated HI abruptly reaching its upper or lower limits, subsequently impacting the LCC count in the subsequent interval. To avert the oscillation

of HI due to such positive feedback, the project introduces a judicious smoothing buffer mechanism during the HI calculation process. This mechanism utilizes the weighted average WLCC of the LCC across multiple segments as the foundation for HI calculation. To provide further clarity, the calculation formula for WLCC is as follows:

$$WLCC = 0.7*LCC + 0.2*LCC1 + 0.1*LCC2 \tag{1.3}$$

LCC1 represents the LCC value from the previous HELLO transmission interval, while LCC2 denotes the LCC value from the preceding two HELLO transmission intervals. Furthermore, the ALCC value undergoes an update each time the LCC within the HELLO message sending interval is calculated.

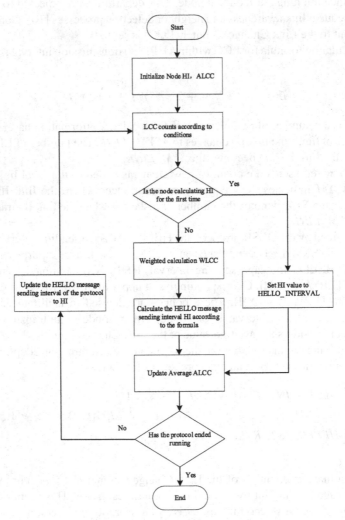

Fig. 1. Processing of sending the HELLO.

The algorithm's process is visually captured in the summarized flow chart depicted in Fig. 1.

2 Analysis of the Whole Process of Demonstration Examples

2.1 Ground End Procedure

(1) UAV relative coordinate flight control terminal my_ send.

The UAV relative coordinate flight control terminal, named "my_send", serves as the publisher node for the ROS topic. This stage involves two message formats: Message1.msg and Message2.msg. Among them, Message1 remains reserved with an unspecified format, while Message2 is the primary focus. The topic is denoted as "nodeX/my_flight_msg". The structure of the message is [float32 x float32 y float32 z int32 flag]. The three float32 values indicate the relative displacement coordinates transmitted to the UAV, measured in meters (m), while the int32 "flag" serves as an identification marker. The sending format follows the pattern of "nodenum x y z flag", where the UAV node label is specified first, followed by the displacement coordinates and identification flag. Typically, the identification flag is set to 1.

For instance, entering "7 24 36 48 1" implies that the UAV node numbered 7 will undergo a movement of 24 m along the X-axis, 36 m along the Y-axis, and 48 m along the Z-axis.

(2) UAV take-off and landing control terminal my_ takeoff.

The UAV takeoff and landing control terminal, known as "my_takeoff", operates as the publisher node for a ROS topic. The topic being published is titled "nodeX/my_takeoff_command", with the message format defined as "takeoff_land.msg", structured as [int32 command]. The commands within this format encompass three categories: "1" signifies "go home", a relatively uncommon and less user-friendly option; "4" indicates "take off"; and "6" represents "land." The sending pattern adheres to "Node command", where the UAV node number is specified first, followed by the control command. For instance, inputting "3 4" denotes that the UAV node labeled as 3 will initiate a takeoff action.

(3) UAV GPS and flight status receiving terminal gps_ rec.

The UAV GPS and flight status receiving terminal, referred to as "gps_rec", operates as an ROS node functioning as a subscriber to specific topics. Its primary task involves receiving data from topics titled "nodeX/dji_sdk/gps_position" and "nodeX/dji_sdk/flight_status." The format of the messages aligns with the sensor data provided by the ROS system, specifically using "msgs::NavSatFix." This format captures information such as longitude, latitude, altitude, and real-time flight status for each individual UAV. Currently, this node discerns the received UAV information and exhibits it on the terminal interface.

(4) UAV plane speed and angular rate control terminal.

The UAV plane speed and angular rate control terminal serves as an ROS topic publisher node responsible for managing the UAV's horizontal and vertical speeds along with its yaw angle. The relevant topic is denoted as "nodeX/dji_sdk/flight_control_setpoint_ENUvelocity_yawrate", and it enables the control of the UAV's angular velocity for yaw, roll, and pitch angles, as well as altitude adjustments. This control is exercised through the topic "nodeX/dji_sdk/flight_control_setpoint_rollpitch_yawrate_zposition." The message formats employed for these two topics utilize "msgs::Joy", sensors provided by the ROS system. The parameters vary as required for each case.

However, due to the current demonstration scheme's lack of clarity regarding the function and objectives of this control terminal, it did not actively participate in the practical demonstration. Instead, it has been reserved as a control interface for potential future use.

2.2 Flight Control Procedure

Utilizing the onboard DJI SDK node, the UAV flight control program establishes a connection between the UAV and the ground control center. This connection facilitates the comprehensive control of UAV takeoff, landing, and flight behaviors. Notably, each UAV is endowed with the capability to function as either a leader or a wingman. This role empowers each UAV to transmit ground-based flight control directives and govern the flight of other UAVs.The broader workflow of the flight control procedure is visually depicted in Fig. 2.

The primary functions and their corresponding roles encompass:

(1) nh.advertise

Use Case: ros:: Publisher ros_ tutorial_ pub =

nh. Advertise < ros_ tutorials_ topic::MsgTutorial > ("ros_tutorial_msg", 100);

Function: create topic publishers for information transmission (sending) between ros nodes.

Remarks: ros_ tutorials_ Topic:: MsgTutorial is the message format, which can be customized; "Ros_tutorial_msg" in the quotation marks is the topic name, and 100 is the cache interval.

(2) nh.subscribe

Use Case: ros:: Subscriber ros_ tutorial_ sub = nh. Subscribe("ros_tutorial_msg", 1000, msgCallback);

Function: create a topic receiver for information transmission (reception) between ros nodes. At the same time, the callback function is used here to trigger when receiving a specific topic.

(3) ros::spin();

Function: When the program runs here, it will wait for the trigger of the callback function to receive the topic continuously.

(4) bool takeoff_ land(int task)

Functions provided by DJI OSDK to control UAV takeoff, landing and return. Task = 1, go home; task = 4, take off; task = 6, land。

(5) bool obtain_ control()

The function provided by DJI OSDK for acquiring control authority of UAV is used to activate Onboard SDK control of UAV, which is the basis of secondary development and flight control program.

(6) void setTarget(float x,float y,float z)

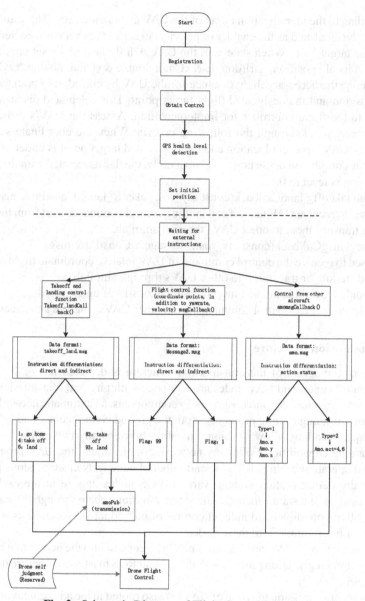

Fig. 2. Schematic Diagram of Flight Control Procedure.

It is used to calculate and set the flight target point position of the UAV. If the z value is too low and negative, it is set to 1.

(7) void local_ position_ callback(const geometry_msgs::PointStamped::ConstPtr& msg)

According to the identification target of the UAV at this time_ set_ The value of state (which can be identified as the enabler or priority) to select the execution content for the UAV. When target_ set_ When state = 1, the UAV will fly toward the set target point.

(8) void local_ position_ ctrl(double &xCmd, double &yCmd, double &zCmd)

Determine the necessary flight distance for the UAV by considering both the UAV's current position and the designated flight target point. This calculated distance is then conveyed to the flight controller for implementation. Assess the UAV's arrival at the intended target point through the following criteria: When the coordinate separation between the UAV's present location and the designated target point is under 10cm, this indicates the completion of the target. Subsequently, the flag associated with this specific flight mission is reset to 0.

(9) void takeoff_ landCallback(const my_ros::takeoff_land::ConstPtr& msg)

It is used to receive UAV takeoff and landing control commands sent from the ground station, or transmit them to other UAVs through amoPub.

(10) void msgCallback(const my_ros::Message2::ConstPtr& msg)

It is used to receive the control command of UAV relative coordinate flight sent from the ground station, or transmit it to other UAVs through amoPub.

(11) void amoCallback(const my_ros::amo::ConstPtr& msg)

It is used to receive control commands from other UAVs, analyze and execute them.

2.3 Transmission Structure

The schematic diagram of the transmission structure is depicted in Fig. 3. Both the ground control center and UAV nodes are equipped with physical transmission boards. Upon powering on the UAV and Raspberry Pi components, an automatic networking process initiates, running routing protocols. Both the ground control center and UAV nodes hold an equal standing, allowing communication between any two entities with relay and forwarding capabilities. Following network establishment, the UAV progresses to initiate and operate the "dji_sdk" node and onboard node tests in succession. Throughout the entire demonstration system, various ROS nodes engage in the exchange of control messages and status updates. These transmissions occur through the ad hoc network, enabling both direct and indirect control of individual UAV nodes, as well as the collection of information from these nodes.

However, when a UAV node becomes newly integrated into the network, it requires a direct or indirect registration process with the ground ROS master node before it becomes controllable.

The video transmission function of UAVs is also rooted in the ad hoc network formed by multiple UAV nodes. A camera affixed to the UAV's Raspberry Pi captures video footage, subsequently transmitting it along the following path: camera → Raspberry Pi → physical transmission board card. The transmitted video is then routed back to the video display section of the ground control center for visual representation. This transmission can occur either through a direct link with the ground control center or through multi-hop forwarding involving other UAV nodes.

QwebEngine offers a method to execute JavaScript within HTML, wherein HTML serves as an interface incorporating < script > < /script > tags. Within these tags, functions and variables are defined, destined either for JavaScript or Qt invocation. Thus,

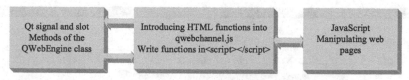

Fig. 3. Qt and Js Call Relationship.

HTML emerges as a bridge interconnecting Qt and web pages. The communication interface is facilitated through the use of "qwebchannel.js."

Simultaneously, the creation of the Baidu online map browser file, named "Map.html", is required. This entails integrating the Baidu Map developer key, determining map dimensions, and adding requisite API functions. Baidu Map offers diverse functional modules encompassing mouse click events, annotations, zooming, display of longitude and latitude coordinates, coverage data, and more.

A pivotal component within this file is the establishment of a new function, "QWebChannel(qt.webChannelTransport, function(channel))", that fosters interaction with Qt. Within this function, a map is created. Event listening is implemented for click events using "map.addEventListener("click")", allowing the invocation of the "bridge_js.getCoordinate()" interface function. This function transfers longitude and latitude coordinates extracted from the interface to Qt using "e.point.lng, e.point.lat".

Within Qt, the creation of a channel object, "channel = new QWebChannel(this)", is essential for interacting with JavaScript. The registered name of this object, "bridge_name", should align with the one in Js.

Ultimately, the channel establishment culminates in writing slot functions using Qt controls, embedding optimization algorithms within these slots. Signal-slot connections enable the encapsulation of flight control algorithms. Upon clicking the designated button, the algorithm executes to calculate UAV flight coordinate data and hover times. The resulting data and track information are displayed on the map interface. The operational ground interface includes UAV position capture, flight cycle configuration, user number specification, optimized trajectory generation, local text file storage, and additional functions. The comprehensive interface and control setup can be observed in the provided Fig. 4.

(1) Overall interface design

The primary debugging procedures entail the following steps: Initially, retrieve the current UAV position by selecting the respective UAV number from the left-hand side of the interface. Proceed to establish the communication cycle on the right-hand side by configuring it through the "Set flight period" button. Additionally, user locations can be designated using the "Set user location" button. Once these parameters are configured, employ the "Generate trajectory" button to generate the trajectory. Subsequently, utilize the "Save to local files" button to store track information and hover times within a local file. The process of trajectory generation and parameter configuration is visually outlined in Fig. 5.

By configuring the UAV's flight cycle to 600 s and identifying 5 users (as indicated by the red arrow in the illustration), the resulting UAV flight path is represented by the red curve. Notably, the optimal hovering position for WPT (Wireless Power Transfer)

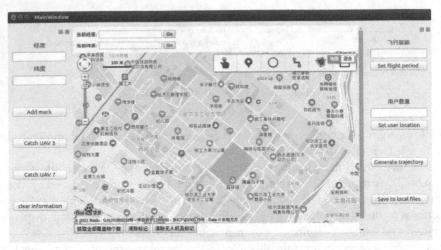

Fig. 4. Main GUI.

Fig. 5. UAV ground end generated trajectory

is illustrated by the red circle. This UAV trajectory data is subsequently saved in a local file. To initiate the process, start the virtual machine terminal and run "roscore" to activate the ROS system. Keep this process running in the background. Then, utilize another terminal to run the my_ Takeoff and my_ Send file. The my_ Send program sends the coordinates and hover time data from the document to the UAV. The UAV subsequently executes the designated tasks in accordance with the specified coordinate sequence. The longitude and latitude coordinates of each UAV are transmitted to the ground control terminal through the Onboard SDK of each UAV via the RoS interface. The map interface dynamically reads and displays the real-time longitude and latitude coordinate information.

Following the completion of software and hardware design, the project confirms the functionality of the ground flight control platform and validates the design algorithm's application using DJI Assistant 2 software provided by DJI Company. This software facilitates flight control command transmission and simulation flight tests on the ground. The "graph" command enables real-time monitoring of the ROS system's operations, as illustrated in Fig. 6.

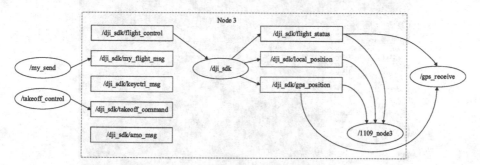

Fig. 6. Operation of ROS Nodes

The data transmitted via our display and control interface first goes through my_ Send and subsequently, the message is sent through ROS to the UAV's flight control system. This orchestration enables the UAV to execute the relevant commands for its flight. The ensuing illustration showcases the simulated flight outcomes, which are transmitted to the UAV terminal. These results are generated using the aforemetioned map to plot the trajectory (Fig. 7).

Fig. 7. Simulated Flight Path of UAV

As evident from the depicted figure, the UAV's flight trajectory aligns seamlessly with the path optimized through our algorithm. To offer a more elucidating portrayal of the UAV's course of action, we provide explicit annotations utilizing the top view illustration illustrated in Fig. 8.

From the aerial perspective of the UAV's simulated flight, a distinct pattern emerges as the UAV sequentially traverses five user positions on the map, clearly marked by red

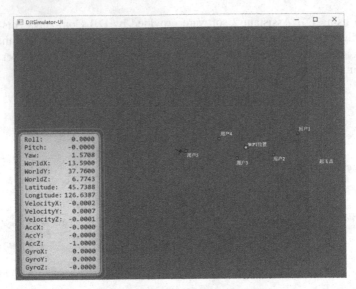

Fig. 8. Top view of UAV simulation track

circles. Additionally, the UAV successfully undergoes charging at the Wireless Power Transfer (WPT) hovering position, distinctly indicated by yellow stars. With our algorithm optimizing the hover times, the UAV communication cycle is set to 600 s, ensuring that it hovers above the designated user and WPT locations at the specified intervals. During its hover over users, it transmits in Wireless Information Transfer (WIT) mode. The simulation outcomes underscore the tangible real-world utility of our algorithm.

Subsequent to the ground-based simulation testing, the project progressed to an actual outdoor flight test. Leveraging the DJI Guidance navigation system, the UAV demonstrates the capability to autonomously navigate around obstacles. Through this real-world flight test, the project successfully validated the algorithm's feasibility and practical applicability. The results of the actual flight test are depicted in the Fig. 9.

Fig. 9. Actual Flight Test UAV

References

1. Zeng, Y., Zhang, R., Lim, T.J.: Throughput maximization for UAV-enabled mobile relaying systems. IEEE Trans. Commun. **64**(12), 4983–4996 (2016). https://doi.org/10.1109/TCOMM.2016.2611512
2. Hua, M., Wang, Y., Zhang, Z., Li, C., Huang, Y., Yang, L.: Outage probability minimization for low-altitude uav-enabled full-duplex mobile relaying systems. China Commun. **15**(5), 9–24 (2018)
3. Zhang, S., Zhang, H., He, Q., Bian, K., Song, L.: Joint trajectory and power optimization for uav relay networks. IEEE Commun. Lett. **22**(1), 161–164 (2018)
4. Zhan, P., Kai, Y., Swindlehurst, A.L.: Wireless relay communications with unmanned aerial vehicles: performance and optimization. Aerospace Electron. Syst. IEEE Trans. **47**(3), 2068–2085 (2011)
5. Khuwaja, A.A., Zheng, G., Chen, Y., Feng, W.: Optimum deployment of multiple UAVs for coverage area maximization in the presence of co-channel interference. IEEE Access **7**, 85203–85212 (2019)
6. Mozaffari, M., Saad, W., Bennis, M., Debbah, M.: Efficient deployment of multiple unmanned aerial vehicles for optimal wireless coverage. IEEE Commun. Lett. **20**(8), 1647–1650 (2016)
7. Jiang, F., Swindlehurst, A.L.: Optimization of UAV heading for the ground-toair uplink. IEEE J. Sel. Areas Commun. **30**(5), 993–1005 (2012)
8. Challita, U., Saad, W., Bettstetter, C.: Interference management for cellular connected UAVs: a deep reinforcement learning approach. IEEE Trans. Wireless Commun. **18**(4), 2125–2140 (2019)
9. Mei, W., Zhang, R.: Uplink cooperative noma for cellular-connected uav. IEEE J. Selected Top. Signal Process. **13**(3), 644–656 (2019)
10. Mozaffari, M., Saad, W., Bennis, M., Debbah, M.: Unmanned aerial vehicle with underlaid device-to-device communications: performance and tradeoffs. IEEE Trans. Wireless Commun. **15**(6), 3949–3963 (2016)

A Lightweight Fault Diagnosis Model of Rolling Bearing Based on Gramian Angular Field and EfficientNet-B0

Yingyu Dai[1], Jingchao Li[1(✉)], Yulong Ying[2], Bin Zhang[3], Tao Shi[4], and Hongwei Zhao[4]

[1] College of Electronic and Information Engineering, Shanghai Dianji University, Shanghai, China
lijc@sdju.edu.cn
[2] School of Energy and Mechanical Engineering, Shanghai University of Electric Power, Shanghai, China
[3] Department of Mechanic Engineering, Kanagawa University, Yokohama, Japan
[4] Shandong Baimeng Information Technology Co., Ltd., Weihai, China

Abstract. The deep learning model can fully extract the rich information in the signal and provide a better recognition effect as compared to traditional fault diagnosis approaches, which rely on manual analysis. However, the deep learning model has the problem that too many parameters lead to high training cost and low generalization ability. Therefore, this paper proposes a lighter fault diagnosis model for rolling bearings using Gramian Angular Field (GAF) and EfficientNet-B0. Firstly, Gramian Angular Field is used to encode one-dimensional vibration signal into a two-dimensional temporal image, the two-dimensional image is then loaded into the selected EfficientNet-B0 for training in automatic feature extraction and classification recognition before a test set is used to confirm the model's recognition accuracy. The results show that the recognition rate of bearing faults of the lightweight fault diagnosis model proposed in this paper based on Gramian Angle field and EfficientNet-B0 reaches 99.27%, and the number of parameters of the EfficientNet-B0 is about 1/5 of that of the ResNet-50. Compared with common fault diagnosis methods, this model has a lighter weight convolutional network model, better generalization and higher recognition rate.

Keywords: Gramian Angular Field · EfficientNet-B0 · intelligent fault diagnosis · time series imaging

1 Introduction

Bearings are one of the most widely used mechanical products in machinery and equipment. As an important part of rotating machinery, its failure will

Supported by Shanghai Rising-Star Program(No.23QA1403800), National Natural Science Foundation of China (No.62076160) and Natural Science Foundation of Shanghai (No. 21ZR1424700).

J. Li et al. (Eds.): 6GN 2023, LNICST 554, pp. 188–199, 2024.
https://doi.org/10.1007/978-3-031-53404-1_16

have an important effect on the performance and stability of the equipment [1]. Therefore, timely and accurate fault diagnosis of bearings has significant practical implications.

The original vibration signal of a rolling bearing contains a large amount of rich information, and the analysis of the vibration signal of the bearing has also become an important tool for bearing fault diagnosis. Time domain statistical analysis, Fourier spectrum analysis, wavelet transform, empirical modal decomposition and other methods are usually used to extract the representative time and frequency domain features of the original signal [2], such as energy entropy. Sun et al. obtained the time-frequency matrix of variable speed fault signals by fusing time-frequency analysis, and used an improved dynamic path planning method to extract the speed profile to interpolate and resample the time-varying signals in order to realize the fault diagnosis of rolling bearings [3]. Yao et al. proposed a bearing fault diagnosis method using wavelet packet transform for noise cancellation processing and extraction of signal features and classification identification by an extreme learning machine [4]. Ma et al. used variable modal decomposition combined with Teager energy spectrum to achieve more accurate fault diagnosis [5].

With the increase of the number of monitoring points of mechanical equipment, the difficulty of identifying bearing fault types is also increased. At this time, it is hard to achieve ideal diagnosis effect by using traditional fault diagnosis technology. In this case, the intelligent diagnosis theory based on deep learning attracts more extensive research, because the theory can adapt to the changes and uncertainties of the signal, and better deal with nonlinear and nonstationary signals, which is suitable for solving such problems. Among them, the convolutional neural networks(CNNs) with adaptive feature extraction ability, is frequently used in the intelligent fault diagnosis. Li et al. proposed a model combining short-time Fourier transform and convolutional neural network for fault diagnosis of rolling bearings [6]. The parameter sharing and local perceptibility of convolutional neural networks make the models more flexible and show better results for feature extraction and recognition tasks on two-dimensional images. Determining how to transform the original one-dimensional vibration signal into two-dimensional picture data for input into a convolutional neural network for fault diagnostics has therefore also become a new study direction.

Although the above research has achieved good diagnostic results, there are still the following problems: 1) the bearing vibration signal has obvious time characteristics, and the traditional means of analysing the bearing vibration signal will cause a certain amount of information loss; 2) The parameters of many convolutional neural networks currently used are relatively complex, and there are problems such as high training cost and low generalization ability, so the feature recognition of fault states is not ideal. In this paper, based on the characteristics of existing methods, we propose a diagnosis method using Gramian Angular Field and EfficientNet-B0. We use Efficientnet-B0 to conduct an experiment, which is the smallest model in the EfficientNet model. The original bearing vibration signals are converted using Gramian Angular Field, and the images thusly encoded are then sent into the EfficientNet-B0 for feature extraction and classification recognition automatically. The Gramian Angular Field technology retains the fault characteristics more completely [7], and the EfficientNet model features fewer parameters, lighter weight and better generalization capability, effectively improving the bearing fault diagnosis efficiency.

2   Time Series Imaging

Computer vision technology is constantly developing, and now people are beginning to study how to use computer vision technology to classify raw vibration signals. The process of converting the original vibration signal into a two-dimensional image (or array), which needs to display as many features and patterns of the original vibration signal as possible, is crucial [7].

In this paper, the one-dimensional signal sequence is encoded as an image using Gramian Angular Field, which represents the disturbance signal sequence in polar coordinates rather than in the typical Cartesian coordinate system. The specific coding process is as follows: given a one-dimensional disturbed signal, $X = \{x_1, x_2, \cdots, x_n\}$, in which the time series consists of n time stamps t , and corresponding actual observed values x, the time series is scaled to [-1,1] by Equation (1), and then the value is encoded into Angle cosine and the time stamp is taken as the radius r. Finally, Equation (2) is used to convert X into a point in polar coordinates. The time series $\tilde{X}$ is obtained.

$$\tilde{x}_i = \frac{(x_i - \max(X)) + (x_i - \min(X))}{\max(X) - \min(X)} \tag{1}$$

$$\begin{cases} \phi = \arccos(\tilde{x}_i), -1 \leq \tilde{x}_i \leq 1, \tilde{x}_i \in \tilde{X} \\ r = \frac{t_i}{N}, t_i \in N \end{cases} \tag{2}$$

Where, t_i is the time stamp, and the interval [0,1] is divided into N equal parts, which is the time series is normalized on the time axis, regularizing the span of the polar coordinate system [8]. Because $\cos(\phi)$ monotonically decreases at $\phi \in [0,\pi]$, the encoded mapping strategy of Equation (2) ensures bijectivity. Each point in the polar coordinate system has a unique timestamp and value given a time series, and in addition its inverse mapping is unique. This encoding is different from the encoding in Cartesian coordinates, in polar coordinates the magnitude of the timestamp t_i is still dependent on the time series, so it maintains an absolute time relationship.

Will be mapped to one dimensional signal after the polar coordinates, the Angle of the polar coordinates of each point can be seen as a timestamp, we can according to the perspective to identify different time intervals of time dependence. The angular information in the polar co-ordinate system can also be used to capture phase information in the signal. Thus, by mapping the signal into a polar co-ordinate system, we can better understand the relationship between the temporal correlation and the frequency component of the signal. Depending on the trigonometric functions used in the encoding process, the encoding method of the Gramian Angular Field can transform the time series into 2 characteristic images, namely the Gram's angular difference field (GADF) map based on the sine function and the Gram's angular sum field (GASF) map based on the cosine function, which are defined in Eqs. (3) and (4) [9].

$$G_{GADF} = \begin{bmatrix} \sin(\phi_1 - \phi_1) & \cdots & \sin(\phi_1 - \phi_n) \\ \sin(\phi_2 - \phi_1) & \cdots & \sin(\phi_2 - \phi_n) \\ \vdots & & \vdots \\ \sin(\phi_n - \phi_1) & \cdots & \sin(\phi_n - \phi_n) \end{bmatrix} = \sqrt{I - \left(\tilde{X}^2\right)^T} \cdot \tilde{X} - \tilde{X}^T \cdot \sqrt{I - \tilde{X}^2}$$

(3)

$$G_{GASF} = \begin{bmatrix} \cos(\phi_1 - \phi_1) & \cdots & \cos(\phi_1 - \phi_n) \\ \cos(\phi_2 - \phi_1) & \cdots & \cos(\phi_2 - \phi_n) \\ \vdots & & \vdots \\ \cos(\phi_n - \phi_1) & \cdots & \cos(\phi_n - \phi_n) \end{bmatrix} = \tilde{X}^T \tilde{X} - \sqrt{I - \left(\tilde{X}^2\right)^T} \cdot \sqrt{I - \tilde{X}^2}$$

(4)

Where, I is the unit row vector; $\tilde{X}^T$ is the transpose vector of $\tilde{X}$. The pixel points in the feature map correspond to different locations and timestamps, and the spatial relationships between neighbouring pixel points are similar to those between the corresponding local areas in the original data, thus preserving the spatial correlation in the original data. The temporal correlation of the original data remains the same, while the time increases when changing from the top left corner to the bottom right corner (Figs. 1 and 2).

Fig. 1. GADF image

Fig. 2. GASF image

The Gramian Angular Field uses the gradient vector of each pixel in the image to describe the characteristics of the point, and then forms a matrix G, which is called Gram matrix. The definition is shown in Eq. (5) [8–10].

$$G = \begin{bmatrix} \langle \tilde{x}_1, \tilde{x}_1 \rangle & \cdots & \langle \tilde{x}_1, \tilde{x}_n \rangle \\ \langle \tilde{x}_2, \tilde{x}_1 \rangle & \cdots & \langle \tilde{x}_2, \tilde{x}_n \rangle \\ \vdots & & \vdots \\ \langle \tilde{x}_n, \tilde{x}_1 \rangle & \cdots & \langle \tilde{x}_n, \tilde{x}_n \rangle \end{bmatrix}$$

(5)

The values on the main diagonal are highly correlated because they correspond to adjacent time points, and the high-level features that the deep neural network learned can capture the significant features in the time series. Using the major diagonal, the time series can be approximately reconstructed in accordance with the high-level features of deep neural network learning.

3   EfficientNet-B0 Model

3.1   EfficientNet-B0 Model Network Structure

The EfficientNet model is an efficient and high-performance new lightweight convolutional neural network architecture proposed by Google researchers in 2019 that excelled on the ImageNet dataset, achieving the best accuracy at the time. The EfficientNet model uses a simple but efficient composite coefficient to uniformly scale all dimensions of depth, width and resolution, achieving very ideal accuracy with fewer parameters and on the order of FLOPS [11]. The main building block is the inverted bottleneck MBConv, SE (extrusion and incentives) optimization is added in it. MBConv is a lightweight convolution operation that reduces computational effort by splitting the standard convolution into two steps: depth convolution and point-by-point convolution. Deep convolution applies the convolution kernel independently on each input channel, and then integrates by applying 1×1 point-by-point convolution on the output channel. Deep separable convolution can reduce computation by a factor of nearly k^2 when compared to traditional convolution, where k is the kernel size and denotes the width and height of the two-dimensional convolution window.

The scaling factor of the EfficientNet model is composed of composite factors, including depth scaling factor d, width scaling factor w and resolution scaling factor r. These scaling factors can adjust the network structure adaptively to different computing resources and task requirements. The EfficientNet model uses a composite coefficient φ to uniformly scale the network width, depth and resolution, as shown in Eq. (6).

$$\begin{cases} depth : d = \alpha^\varphi \\ width : w = \beta^\varphi \\ resolution : r = \gamma^\varphi \end{cases}, \alpha \geq 1, \beta \geq 1, \gamma \geq 1 \qquad (6)$$

Where φ is a user-specified factor and defines how many resources are available for model scaling and a grid search can be used to find α, β, γ, which is a constant. These extra resources are distributed to network breadth, depth, and resolution, respectively, according to α, β, γ. In regular convolution process, FLOPS is proportional to d, w^2, r^2, meaning that FLOPS doubles as the depth of the network doubles and quadruples as the width or resolution of the network doubles.

We apply our composite scaling method to scale it in two steps and start with the baseline EfficientNet-B0:

Step 1: Assuming twice the available resources, perform a grid search with $\varphi=1$ to find the best values of α,β,γ.

Step 2: Determine the acquired values of α,β,γ as constants. Using Equation (6), amplify the baseline network with various values of φ to yield EfficientNet-B1 to B7 [11,12].

The EfficientNet-B0 model has a parametric count of 5.3M, about 1/5 of the ResNet-50 model, and a FLOPS of about 1/11 of the ResNet-50 model. These results indicate that EfficientNet has higher model efficiency and better utilization of computational resources [11–13] (Table 1).

Table 1. Comparison of different model parameter numbers and FLOPS.

Model	parameters	Ratio-to-EfficientNet	FLOPS	Ratio-to-EfficientNet
EfficientNet-B0	5.3M	1x	0.39B	1x
ResNet-50	26M	4.9x	4.1B	11x

By taking different values of φ, a series of EfficientNet models of different sizes can be constructed. In this paper, the EfficientNet-B0 is used. The model network results are shown in the following table [11] (Table 2):

Table 2. Network structure of the EfficientNet-BO.

Stage i	Operator Fi	Resolution Hi ×Fi	Channels Ci	Layers Li
1	Conv3×3	224×224	32	1
2	MBConv1,k3×3	112×112	16	1
3	MBConv6,k3×3	112×112	24	2
4	MBConv6,k5×5	56×56	40	2
5	MBConv6,k3×3	28×28	80	3
6	MBConv6,k5×5	14×14	112	3
7	MBConv6,k5×5	14×14	192	4
8	MBConv6,k3×3	7×7	320	1
9	Conv1×1 & Pooling & FC	7×7	1280	1

3.2   Transfer Learning

Transfer learning is a machine learning technique that applies knowledge gained while training on one issue to training in another task or domain. Training deep learning models is expensive and requires large data sets, so transfer learning is often applied to the field of deep learning due to its effectiveness [14].

From different areas of the image contains the underlying characteristics of common, so you can use in the source area to learn the underlying characteristics of convolution neural network to help in the field of target image classification task. Transfer learning through the migration of common characteristics in the convolution layer knowledge, make learning more stable, so as to improve the training efficiency. It uses the parameters of the pre-trained model as initial values, avoids the problem of random initialization of the training model, makes the model converge faster, and reduces the possibility of overfitting. Because the pre-trained model has already learned many common features, by fine-tuning the model, it is possible to further optimize the model to adapt to the new data set while retaining the common features. This has significant advantages over training the model from scratch in terms of saving time and achieving highly accurate identification [15, 16].

The study trained the Efficientnet-B0 network based on transfer learning as follows: we can use trained weights because the Efficient network model has already been trained on the ImageNet dataset, which is a large public dataset. This is then combined with the Gramian Angular Field encoded training set to improve the learning performance of the network model on the dataset by fine-tuning the network model parameters through optimisation learning. Using a model already trained on one dataset and applying it to a new dataset or task decreases the number of training parameters required, reduces the risk of over-fitting, generalises and processes new data better and speeds up the convergence of the model.

4   Simulation and Analysis

4.1   Data Sources and Processing

This paper used the publicly available bearing dataset from Case Western Reserve University, whose primary model is an end deep groove ball bearing using a motor drive model SKF 6205, with damage to the bearing being single point damage machined with EDM. The test bench consists of a horsepower motor, torque sensor/encoder, force meter and control electronics, and the vibration signal comes from a 16-channel data logger, as shown in Fig. 3.

Fig. 3. Fault simulation test rig for rolling bearings.

In this paper, 48kHz bearing signal sampling frequency in Case Western Reserve University dataset was selected, motor speed was 1797r/min, the fault diameters include 0.007 in., 0.014 in., and 0.021 in., the fault location is respectively in the ball, inner raceway, outer raceway, in addition to the normal state, a total of 10 fault modes were selected, as shown in Table 3.

Table 3. Description of data set classification.

Fault location	Diameter/inch	Tags
Ball	0.007	0
Ball	0.014	1
Ball	0.021	2
Inner Raceway	0.007	3
Inner Raceway	0.014	4
Inner Raceway	0.021	5
Outer Raceway	0.007	6
Outer Raceway	0.014	7
Outer Raceway	0.021	8
Normal	None	9

The use of sliding window sampling is a method of splitting a signal in a data set using a fixed-length sliding window, allowing more samples to be obtained in a limited data set. The sliding window's length should be set to the sample's length, which was 2000. The sliding window's moving step was set to one move's distance of distance, and the moving step was set to 1000, so that there was partial overlap between the two adjacent samples, and the overlap rate was 50%, so that more time data samples are collected for the model's training and testing. Through sliding window sampling, 480 time images are obtained for each bearing state, achieving a 6:2:2 ratio between the training set, verification set and test set.

4.2   Analysis and Comparison of Training Results

After encoding the 1D vibration signal using Gramian Angular Field, in order to train the EfficientNet-B0 utilizing transfer learning, the generated GADF and GASF image training sets were fed into it. The constructed training set and verification set were used to conduct fault diagnosis simulation and comparison on two different models. Illustrations below display the simulation results for the two models. At the same time, in order to evaluate the benefits and drawbacks of the proposed model more clearly and easily, the confusion matrix was also drawn after model training, so as to assess the effectiveness of the classification model (Figs. 4, 5, 6, 7, 8 and 9).

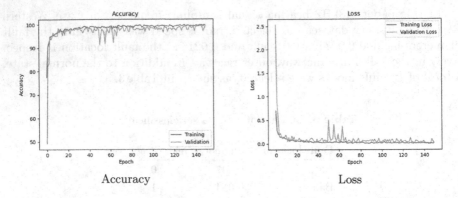

Fig. 4. EfficientNet-B0 input GADF image simulation results.

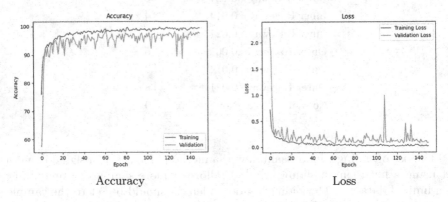

Fig. 5. EfficientNet-B0 input GASF image simulation results.

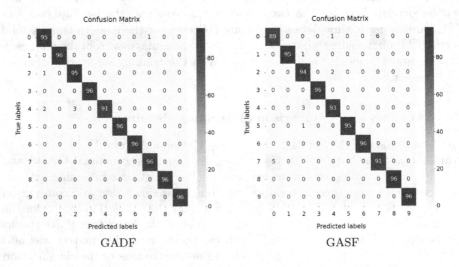

Fig. 6. Confusion matrix.

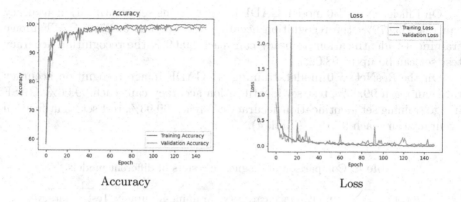

Accuracy   Loss

Fig. 7. ResNet-50 input GADF image simulation results.

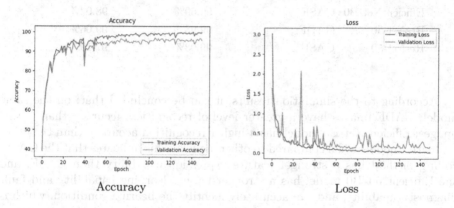

Accuracy   Loss

Fig. 8. ResNet-50 input GASF image simulation results.

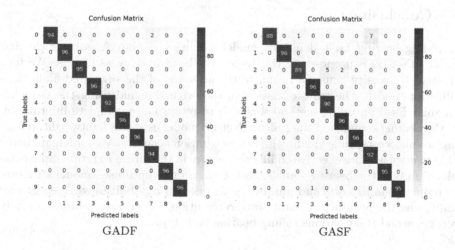

GADF   GASF

Fig. 9. Confusion matrix.

On EfficientNet - B0 model, GADF image training set identification accuracy can reach 99.79%, the recognition accuracy test set was 99.27%, GASF image training set identification accuracy can reach 99.69%, the recognition accuracy test set can be up to 98.02%.

On the ResNet - 50 model, training set GADF image recognition accuracy rate can reach 99.93%, test set identification accuracy can reach 99.06%, GASF image training set identification accuracy can reach 99.93%, test set identification accuracy can reach 97.19% (Table 4).

Table 4. Comparison of diagnostic results of different models.

Model	Input image category	Training accuracy	Test accuracy
EfficientNet-B0	GADF	99.79%	99.27%
EfficientNet-B0	GASF	99.69%	98.02%
ResNet-50	GADF	99.93%	99.06%
ResNet-50	GASF	99.93%	97.19%

According to the simulation results, it can be concluded that: on the same model, GADF images have a greater level of recognition accuracy than GASF images; EfficientNet-B0 model has a higher recognition accuracy than the other evaluated models when compared to other models. This indicates that the GADF coding method has a stronger feature expression capability than GASF, and the EfficientNet-B0 model has a stronger feature learning capability and fault diagnosis capability, and can accurately identify the bearing condition, which is better than other comparative models.

5   Conclusions

A rolling bearing fault diagnosis model usining Gramian Angular Field and EfficientNet-B0 is proposed in this paper, the bearing data set of Case Western Reserve University is used to confirm the accuracy of the suggested fault identification model. The transformation of the original signal using the Gramian Angular Field encoding method was able to retain more complete information in the bearing signal, and the EfficientNet-B0 model used was lightweight, reducing training costs while yielding higher recognition accuracy. We anticipate being able to employ more effective coding techniques in the future to retain more complete faults in the bearing signals, and to use a more optimised model structure to reduce training costs and improve generalisation capabilities while effectively improving the accuracy of fault state recognition, in order to more accurately recognize and treat various rolling bearing fault types.

References

1. Zhang, L., Hu, Y., Zhao, L., Zhang, N., Wang, X., Wen, P.: Fault diagnosis of rolling bearings using recursive graph coding technique and residual network. J. Xi'an Jiaotong Univ. **57**(02), 110–120 (2023)
2. Liu, R., Yang, B., Zio, E., et al.: Artificial intelligence for fault diagnosis of rotating machinery: a review. Mech. Syst. Signal Process. **108**, 33–47 (2018)
3. Xinwei, S., Ji Aimin, D., Zhantao, C.X., Xinhai, L.: Diagnosis method of variable speed fault of rolling bearing in gearbox of rolling stock. J. Harbin Instit. Technol. **55**(01), 106–115 (2023)
4. Fenglin, Y., Changkai, X., Shining, L., Hao, Y., Zhe, M.: Research on rolling bearing fault diagnosis based on wavelet packet transform and ELM. J. Saf. Environ. **21**(06), 2466–2472 (2021). https://doi.org/10.13637/j.issn.1009-6094.2020.0999
5. Qiang, M., Yachao, L., Zheng, L., Zhaojian, G.: Fault feature extraction of rolling bearings based on variational modal decomposition and Teager energy operator. Vibration and Shock **35**(13), 134–139 (2016). https://doi.org/10.13465/j.cnki.jvs.2016.13.022
6. Heng, L., Hydrogen, Z., Xianrong, Q., Yuantao, S.: A bearing fault diagnosis method based on short-time Fourier transform and convolutional neural network. Vibr. Shock **37**(19), 124–131 (2018). https://doi.org/10.13465/j.cnki.jvs.2018.19.020
7. Sun, X., Wang, M., Zhan, B., et al.: An intelligent diagnostic method for multi-source coupling faults of complex mechanical systems. Shock and Vibration (2023)
8. Zheng, W., Lin, R.Q., Wang, J., Li, Z.J.: Power quality disturbance classification based on GAF and convolutional neural network. Power System Protect. Control **49**(11), 97–104 (2021). https://doi.org/10.19783/j.cnki.pspc.200997
9. Yao, L., Mianjun, S., Chenbo, M.: A rolling bearing fault diagnosis method based on Gram's angular field and CNN-RNN. Bearings (02), 61–67 (2022). https://doi.org/10.19533/j.issn1000-3762.2022.02.012
10. Han, B., Zhang, H., Sun, M., et al.: A new bearing fault diagnosis method based on capsule network and Markov transition field/Gramian angular field. Sensors **21**(22), 7762 (2021)
11. Tan, M., Le, Q.: EfficientNet: rethinking model scaling for convolutional neural networks. In: International Conference on Machine Learning, pp. 6105–6114. PMLR (2019)
12. Atila, U., UcSar, M., Akyol, K., et al.: Plant leaf disease classification using EfficientNet deep learning model. Ecol. Inform. **61**, 101182 (2021)
13. Li, W., Zhu, X., Gong, S.: Person re-identification by deep joint learning of multi-loss classification. arXiv preprint arXiv:1705.04724 (2017)
14. Weiss, K., Khoshgoftaar, T.M., Wang, D.D.: A survey of transfer learning. J. Big Data **3**(1), 1–40 (2016)
15. Yu, G., Qingwen, G., Chuntao, W., et al.: Crop pest identification based on improved EfficientNet model. J. Agric. Eng., 038-001 (2022)
16. Shorten, C., Khoshgoftaar, T.M.: A survey on image data augmentation for deep learning. J. Big Data **6**(1), 1–48 (2019)

ISAC Device-Free Sensing Method for V2V System

Feiqiao Yu[1], Zhimin Chen[1,2(✉)], and Minzheng Li[1]

[1] The School of Electronic and Information, Shanghai Dianji University,
Shanghai 201306, China
[2] The Department of Electronic and Information Engineering,
The Hong Kong Polytechnic University, Hong Kong, China
chenzm@sdju.edu.cn

Abstract. With the development of information technology, integrated sensing and communication (ISAC) can be directly used in the vehicle-to-vehicle (V2V) scenarios. Since the ISAC system can reduce the communication overhead and realize the device-free localization of vehicles in noncooperative scenarios. A novel passive vehicle location method combines the Doppler time-sharing and the Barker code is proposed in this paper. By using the proposed method suppress the clutter interference, the target vehicles can be accurately located separately. The simulation results show that we can finally realize the accurate perception of 6 vehicles within a kilometer range.

Keywords: integrated sensing and communication ·
vehicle-to-vehicle · device-free · suppress the clutter interference

1 Introduction

Due to the rapid growth of wireless technologies like the Internet of Things and intelligent driving, wireless communication technology faces increasing performance demands. However, current wireless communication systems are struggling to keep up with the complexity of the wireless environment. ISAC technology has gained attention as a hot research topic due to its advantages, including efficient use of frequency spectrum.

Device-free sensing, as a key research direction, enables radar detection through signals reflected by targets [1]. This approach finds valuable applications in vehicle networking. In vehicle networking, we utilize device-free sensing technology for radar-assisted communication, effectively reducing communication costs and enhancing communication reliability [2]. Moreover, vehicles can

Supported by Shanghai Natural Science Foundation project (No. 22ZR1425200) and Science and Technology Project of Shanghai Science and Technology Commission (No. 21010501000).

continuously detect the surrounding environment through excellent perception technology, obtaining accurate perception information, and exchanging information with neighboring vehicles and roadside units.

Numerous studies have explored the integration of sensors into communication systems [3,4]. One approach involves the vehicle estimating its state by receiving pilot signals from the road side unit (RSU) and providing feedback to improve communication quality. However, the use of pilot frequencies and frequent feedback can result in high communication costs and delays. To address these issues, recent research has focused on extracting vehicle state information from reflected sensing echoes, which offers the advantage of low overhead [5,6]. The authors of [7] compare the sensed vehicle position with the position provided by GPS (Global Positioning System) using an algorithm. However, this approach is hindered by ineffective communication when the GPS signal strength is weak. Consequently, a vehicle perception system relying on satellite positioning or base stations can be influenced by external factors. In this paper, we directly implement equipment-free vehicle sensing in the V2V communication, significantly enhancing the vehicles' ability to process echo information and eliminating dependence on external factors for perception.

This paper aims is to extract the required distance information from clutter signals. The signal transmission on the vehicle employs a pulse-Doppler radar operating in the millimeter-wave frequency range, which provides strong clutter suppression capabilities for the vehicle's radar system, enabling it to distinguish target echoes from strong clutter backgrounds. When the radar signal contacts the vehicle, the embedded communication information in the radar signal is received by the target and a portion of the radar signal is reflected. However, the received echo signal is contaminated with clutter signals. In this paper, we apply Doppler processing to the echo signals and detect the desired vehicle distance information using Barker codes.

The remainder of this paper is organized as follows. Section 2 models the experiment and analyzes the experimental process. Section 3 provides a theoretical introduction to the techniques and methods employed in the experiment. Section 4 presents the simulation results of this experiment, and Sect. 5 concludes the paper.

2   System Model

Pulse radar is a common radar used for sensing and ranging. In the work, we will leverage the advantages of communication sensing integration to enhance the range resolution of the entire system. In general, the communication awareness integration expression can be expressed as

$$S(t) = \alpha s(t - \tau) + n(t) \tag{1}$$

where $S(t)$ is the received ISAC signal, α is the attenuation factor of radar echo signal, $s(t - \tau)$ is the echo signal, $n(t)$ is the clutters.

In this work, the number of target vehicles is set as 6, with the radar reflectivity of the vehicles assumed to be constant by default. However, in real-world scenarios, vehicles can vary in size and shape, leading to different radar reflectivity for each vehicle. Moreover, for the detection of these six moving targets, to better simulate real-world applications, it was decided to generate the range and speed of each target using a random function. Once the communication perception-integrated radar signal is reflected by the target, the combined echo signal from each target can be expressed as

$$s(t,m) = \sum_{k=1}^{6} \rho_k b(t-\tau)e^{jk_0u_kmT} \tag{2}$$

where t is the fast processing time, m is the slow time pulse index, $b(t)$ is the transmitted signal, k_0 is the radar propagation constant, ρ_k is the reflectivity of different vehicles, T is the pulse repetition period, τ is the delay. We can also find the echo of each target by using this equation, and define the delay of the echo signal as $\tau = \frac{2R}{c}$. Then the formula can be rewritten as

$$s(t,m) = \sum_{k=1}^{6} \rho_k b(t-\frac{2R_k}{c})e^{jk_0u_kmT} \tag{3}$$

Then, we received and processed the echo signal of millimeter wave pulse radar. Considering that radar signal would be doped with clutter signal on the way back, we modeled the clutter signal as

$$c = \sqrt{-2\log(1-rand(1,1000)/SCR\,)}e^{2\pi rand(1,1000)j} \tag{4}$$

We combine the clutter signal with the radar's echo signal to obtain the received echo signal at the radar receiver. Then, we compress the echo signal using Barker code, which introduces phase shifts to increase the bandwidth and prevent resolution loss. For the compressed radar signal, we employ a two-dimensional time-sharing processing approach, similar to the accumulation method used in frequency-modulated continuous-wave (FMCW) radar. This involves performing time-sharing processing on the one-dimensional signal, dividing it into fast time and slow time for detailed processing, as shown in Fig. 1.

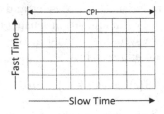

Fig. 1. 2D time-sharing processing

In this work, we construct a two-dimensional array matrix consisting of 1024 range bins. The fast time dimension represents a sequence of pulse measurements within the same range unit. Each column in the two-dimensional data matrix, known as the slow time dimension, corresponds to the continuous sampling of a pulse echo, referred to as a continuous range bin. During the fast processing stage, we sample an additional 1000 echo signal samples. For each sample, an independent clutter value is generated. We then construct a matching filter using the reverse Barker code coefficients. The echo samples are processed through the matching filter and loaded into the respective range bins. In the subsequent slow processing stage, we perform a Fourier transform on the 1000 samples. This Fourier transform serves to transform the signal representation from the slow time domain to the Doppler frequency domain, and from the fast time domain to the range domain. The resulting transformed data is depicted in Fig. 2.

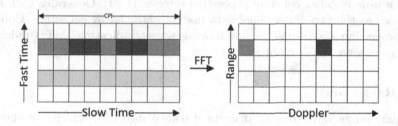

Fig. 2. Slow processing stage

Although the target's distance is generally assumed to remain relatively constant within a short time period, even for moving targets, changes in signal phase should not be overlooked. Even slight variations in distance or time can lead to significant phase changes, which can be utilized for velocity measurement. Rather than employing filtering processing, spectrum analysis is directly applied to the slow time data series. The resulting processed data is organized in a matrix format, with the coordinates transformed into fast time and Doppler frequency.

In the final step, multiple filtered echoes are combined or stacked to generate the clutter-filtered echo signal, which allows for the detection of the target's distance.

3   Principle Analysis

3.1   Time-Sharing Processing

In this work, the echo signal of pulse Doppler radar is processed fast and slow. The principle of fast processing of Doppler radar is to use high sampling rate to measure Doppler frequency shift with high precision. In particular, the fast processing method will continuously collect the complex value of the echo signal in a short period of time and convert it into a frequency domain signal using the

Fourier transform. The Doppler shift of the target can be calculated by analyzing the frequency change in the frequency domain signal. This method is suitable for high-precision Doppler shift measurement, such as high-speed motion or small target detection. Common fast processing algorithms include the fast fourier transform (FFT) and FMCW. The principle of slow Doppler radar processing is to use low sampling rate to measure the Doppler shift roughly. Specifically, the slow processing method picks up the amplitude or phase change of the echo signal over a long period of time and converts it into a direct current (DC) voltage signal using technologies such as phase-locked amplifiers. The Doppler shift of the target can be estimated roughly by measuring the change of DC voltage signal.

In addition, the sampling rate of slow processing is actually the pulse repetition frequency (PRF) of the signal, and we have $PRF = 1/PRI$. If M pulse strings are collected in practice, the time required for data collection is $M \times PRI$, which is usually called coherent processing interval (CPI). Generally, CPI is also expressed as the two-dimensional data matrix obtained by collection. Common slow processing algorithms include moving target indicating (MTI) radar and continuous wave (CW) radar [8].

3.2   Range Bin

In radar, range bin refers to discrete distance intervals, which are classified according to the round-trip time delay of the detected target. This means that the range bin represents a specific distance from the radar antenna, with each bin corresponding to a certain distance interval. The width of each range bin is determined by the pulse repetition frequency (PRF) of the radar system, as shown in Fig. 3.

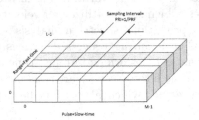

Fig. 3. range bin

Once the radar echo has been sorted into the range bin, various signal processing techniques can be applied to analyze the signal and extract information such as target range, speed and direction. The mathematical formula can be used to explain: suppose the radar distance from the target is R, the flight distance of the radar signal from transmission to return is $2R$, then the flight time is $t = 2R/c$, and when the sampling time is $T_s = 1/f_s$, then the distance the radar travels in time is called the range compartment. Range bin is an important concept in radar signal processing, usually using microwave or millimeter wave such as high frequency signal. If the digital signal processing technology is used to combine multiple range bins into one range unit, the range resolution and measurement accuracy of radar can be improved.

4   Analysis of Experimental Results

This chapter will introduce and analyze the experimental results of simulation. In order to have an intuitive understanding of the experimental results, namely the position perception of the surrounding vehicles, we built a three-dimensional data cube

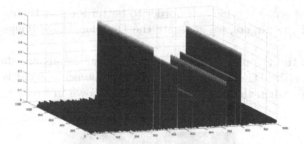

Fig. 4. Data cube

From Fig. 4, we can clearly see the distance between the six vehicles. The horizontal coordinate of Fig. 4 shows the distance between the observation vehicles and the detected vehicles; the vertical coordinate represents the time of the target echo signal received by the observation vehicles; the depth coordinate represents the Doppler frequency in the echo signal. Furthermore, Fig. 5 shows the relationship between the matched filter output and clutter, we can see that, Fig. 5 can also be regarded as a two-dimensional cross-section of Fig. 4.

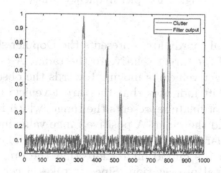

Fig. 5. Data cube

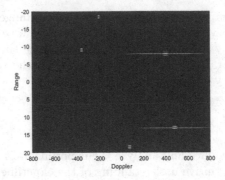

Fig. 6. Doppler frequency-Range

Since the target scene is vehicles, in order to better identify the moving position information of the targets, we generate the Doppler-Range image, as shown in Fig. 6.

However, only the Doppler frequencies of five echo signals are shown in Fig. 6. This is because the Doppler frequency shift of one echo signal is small, which can be observed when the image is enlarged, as shown in Fig. 7.

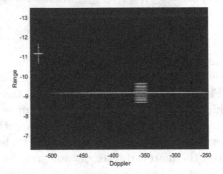

Fig. 7. Doppler frequency-Range

In Fig. 6, the horizontal coordinate represents the Doppler shift, which is a measure of the velocity of the target relative to the radar. A positive Doppler shift indicates that the target vehicle is moving towards the observing vehicle, while a negative Doppler shift indicates that the target vehicle is far away from the radar. The vertical coordinate represents the range, which refers to the distance between the target and the radar. A positive range value indicates that the target is close to the radar, while a negative range value indicates that the target is far from the radar. It's important to note that the range value includes the time delay of the signal propagation. Since it takes a certain amount of time for the radar to receive the echo signal after sending the signal, the range value incorporates this time delay.

In the Doppler-range image, the position of a target can be determined by its Doppler velocity and range distance. Each target in the Doppler-range image appears as a cursor. The size of the cursor corresponds to the strength of the target's echo signal, where a larger cursor indicates a stronger echo signal, while a smaller cursor indicates a weaker echo signal. The strength of the echo signal typically depends on the target's radar cross-section and its distance from the radar.

5   Conclusion

By integrating the pulse Doppler radar with fast and slow processing and utilizing Barker code, the ISAC framework enables effective perception of the positions of nearby vehicles within several hundred meters of the observing vehicle in the new V2V scene. This approach provides a solution to the current overreliance on satellite positioning and base station-assisted positioning for vehicle perception. While this experiment successfully addresses the clutter issue in the echo, further research is required to tackle the challenge of strong interference from the radar transmitter on the receiver signal in the communication perception integration scenario.

References

1. Shi, Q., Liu, L., Zhang, S., Cui, S.: Device-free sensing in OFDM cellular network. IEEE J. Sel. Areas Commun. **40**(6), 1838–1853 (2022)
2. Cheng, X., Zhang, H., Yang, Z., Huang, Z., Li, S., Yu, A.” Research on perceptual integration of internet of vehicles: current situation and development trend. J Commun. **43**(08), 188–202 (2022)
3. Kumari, P., Choi, J., González-Prelcic, N., Heath, R.W.: An approach to joint vehicular communication-radar system. IEEE Trans. Veh. Technol. **67**(4), 3012–3027 (2018)
4. Wymeersch, H., Seco-Granados, G., Destino, G., Dardari, D., Tufvesson, F.: 5G mmWave positioning for vehicular networks. IEEE Wirel. Commun. **24**(6), 80–86 (2017)
5. Liu, F., Yuan, W., Masouros, C., Yuan, J.: Radar-assisted predictive Beamforming for vehicular links: communication served by sensing. IEEE Trans. Wirel. Commun. **19**(11), 7704–7719 (2020)
6. Yuan, W., Liu, F., Masouros, C., Yuan, J., Ng, D.W.K., González-Prelcic, N.: Bayesian predictive beamforming for vehicular networks: a low-overhead joint radar-communication approach. IEEE Trans. Wirel. Commun. **20**(3), 1442–1456 (2021)
7. Szabo, A.P.: Clutter simulation for airborne pulse-Doppler radar. IEEE Cat. No.03EX695, Adelaide, SA, Australia, pp. 608–613 (2003)
8. Maoyong, L., Youan, K.: Radar Signal Theory. National Defence Industry Press, China (1984)

Fuzzy Sliding Mode Trajectory Tracking Control for Omnidirectional Mobile Robots Based on Exponential Convergence Law

Ding Ruiao, Ji Chunlei[⊠], and Zeng Xiangxu

Shanghai Dianji University, Shanghai 201306, China
216003010121@st.sdju.edu.cn, {jicl,zengxx}@sdju.edu.cn

Abstract. The increasing advancements in information technology have led to a growing interest in control research for mobile robot trajectory tracking. A controller for robot systems should exhibit adaptivity and robustness as its fundamental characteristics. Sliding mode control (SMC) offers strong robustness but is hindered by the presence of chattering, limiting its application and development. This paper explores the integration of sliding mode control and fuzzy control to address these limitations. By using the mobile robot's kinematic model, a sliding mode controller based on exponential convergence is designed on the basis of existing sliding mode controllers, incorporating a fuzzy algorithm to mitigate system chattering. Experimental results demonstrate a significant improvement in control performance compared to traditional SMC, with faster convergence of tracking error and enhanced robustness.

Keywords: Omnidirectional mobile robot · Track tracking · Sliding mode control · Lyapunov · kinematic model

1 Introduction

The application of robots has increased in recent years with the rapid development of information technology and artificial intelligence. This development has enabled robots to be used in a variety of scenarios, such as service, exploration, and agriculture. Mobile robots, which are different from traditional industrial robots, are primarily used to explore unknown environments. The main part of the motion control of a mobile robot is tracking control, which involves adjusting the speed and direction of the robot to make it move along the desired trajectory. However, mobile robots are nonlinear systems with multi-variable strong coupling and nonholonomic constraints [1], making trajectory tracking control challenging. Extensive research has been conducted on trajectory tracking control for mobile robots, with common controllers including state feedback linearization control [2], inversion control [3, 4], model predictive control [5, 6], sliding mode control, etc. [7, 8]. Since the omnidirectional mobile robot studied in this paper uses four hub motors to control the forward movement and four servo motors to control the steering, the advantages of sliding mode control for motor control are fully demonstrated compared to other control methods.

J. Li et al. (Eds.): 6GN 2023, LNICST 554, pp. 208–216, 2024.
https://doi.org/10.1007/978-3-031-53404-1_18

Sliding mode control is a robust control scheme with fast transient response. Essentially employing discontinuous non-linear control, it operates in a dynamic process. The advantages of determinism and robustness make sliding mode control useful for mobile robot control. However, the chattering phenomenon inherent to the tracking process can degrade system control performance [9]. Some approaches address this. Literature [10] proposes using intelligent control to estimate the robust part of the sliding mode and eliminate the chattering phenomenon during motion. Additionally, literature [11] proposes a robust controller that combines sliding mode control and backstepping technology, which has a good control effect and can track any trajectory. Literature [12] combines an auxiliary speed controller and an integrated terminal sliding mode controller, performs fuzzy logic approximation on the unknown function, and adds the control strategy of an adaptive fuzzy observer.

Based on the above analysis, combined with the characteristics of the studied omnidirectional mobile robot, this paper firstly establishes a kinematic model of the omnidirectional mobile robot; then designs a sliding mode controller with exponential convergence law to reduce system jitter based on this model; finally, considering the excellent performance of each of fuzzy control and sliding mode control in the control field, combines the two and further builds a fuzzy sliding mode controller, and verifies the stability of the system through The stability of the system is verified by Lyapunov function.

2 Kinematic Modeling of Omnidirectional Mobile Robots

2.1 Kinematic Modeling

The research object of this paper is an omnidirectional mobile robot with independent four-wheel drive and independent steering. Each wheel uses a separate motor to control forward and steering. According to the characteristics of the omnidirectional mobile robot, its model is simplified, and a two-dimensional plane coordinate system is established. The pose error coordinates of the mobile robot are shown in Fig. 1.

In Fig. 1, $(x, y)^T$ represents the current coordinate position of the wheeled mobile robot, θ is the angle between the forward direction and the X-axis, v and ω are the translation speed and rotation speed of the mobile robot, respectively, and they are control input variables in the kinematics model. The nonholonomic constraints of the mobile robot can be expressed by the following formula:

$$\dot{x}\sin\theta - \dot{y}\cos\theta = 0 \tag{1}$$

The kinematic equation of a mobile robot can be described as:

$$\dot{p} = \begin{bmatrix} \dot{x} \\ \dot{y} \\ \dot{\theta} \end{bmatrix} = \begin{bmatrix} \cos\theta & 0 \\ \sin\theta & 0 \\ 0 & 1 \end{bmatrix} q \tag{2}$$

In formula (2), $p = (x, y, \theta)^T$ is the state vector of the mobile robot, let the vector $q = (v, \omega)$ represent the control input.

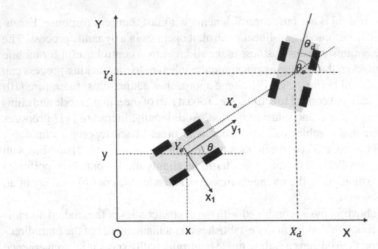

Fig. 1. Kinematics Model of Mobile Robot

2.2 Trajectory Tracking Problem Description

The trajectory tracking problem is to design the control law about v and ω, so that the mobile robot can track the reference trajectory. The sliding mode control process can be described as a control process in which the state trajectory starting from any point in space can reach the sliding mode surface within a certain time, and the sliding mode motion occurs on it, and finally reaches the equilibrium point [13]. The mobile robot moves from pose $p = (x, y, \theta)^T$ to pose $p_d = (x_d, y_d, \theta_d)^T$, the position error of the mobile robot $p_e = (x_e, y_e, \theta_e)^T$, where x_e represents the driving direction error, y_e represents the lateral error, and θ_e represents the direction error.

The geometric relationship shown in Fig. 1, according to the coordinate transformation formula, the error equation of the mobile robot pose can be expressed as:

$$\dot{p}_e = \begin{bmatrix} \dot{x}_e \\ \dot{y}_e \\ \dot{\theta}_e \end{bmatrix} = \begin{bmatrix} y_e\omega - v + v_d \cos\theta_e \\ -x_e\omega + v_d \sin\theta_e \\ \omega_d - \omega \end{bmatrix} \tag{3}$$

From formula (2) and (3), the differential equation for the positional posture can be deduced as:

$$\dot{p}_e = \begin{bmatrix} \dot{x}_e \\ \dot{y}_e \\ \dot{\theta}_e \end{bmatrix} = \begin{bmatrix} y_e\omega - v + v_d \cos\theta_e \\ -x_e\omega + v_d \sin\theta_e \\ \omega_d - \omega \end{bmatrix} \tag{4}$$

After matrix transformation, formula (4) is turned into:

$$\dot{p}_e = \begin{bmatrix} \dot{x}_e \\ \dot{y}_e \\ \dot{\theta}_e \end{bmatrix} = \begin{bmatrix} \cos\theta_e & 0 \\ \sin\theta_e & 0 \\ 0 & 1 \end{bmatrix} \begin{bmatrix} v_d \\ w_d \end{bmatrix} + \begin{bmatrix} -1 & y_e \\ 0 & -x_e \\ 0 & -1 \end{bmatrix} u \tag{5}$$

where $u = \begin{bmatrix} v & \omega \end{bmatrix}^T$, This means that control u is used to solve the tracking problem.

3 Omnidirectional Mobile Robot Controller Design

The trajectory tracking control goal of the mobile robot is to make the pose error e of the mobile robot approach 0 under the control q input, for any initial error, $\lim_{t \to \infty} (x_e, y_e, \theta_e)^T = 0$. The omnidirectional mobile robot in this paper adopts the sliding mode control design of literature [14] and the control law is as follows:

$$\begin{cases} v = v_d \cos \theta_e + y_e w + c x_e + c y_e^2 \\ \omega = \omega_d + \varepsilon \mathrm{sgn}(s) + ks \end{cases} \tag{6}$$

Among them, take the switching function $s = \theta_e$, the exponential approach rate $\mathrm{sgn}(s) = \frac{s}{|s|+\sigma}$, ε, k, and c are all constant and greater than 0.

3.1 Sliding Mode Controller Design Based on Exponential Convergence Law

For a first-order linear system:

$$\dot{x} = u \tag{7}$$

It can be stabilized by the following control law in a finite time, x is the system state variable, $\beta > 0$, m, n are positive odd numbers and satisfy $n > m$.

$$u = -\beta x^{m/n} \tag{8}$$

Substituting formula (8) into formula (7) can get:

$$\frac{dx}{dt} = -\beta x^{m/n} \tag{9}$$

The definite integral of the above formula is obtained from any initial state $x(0) \neq 0$ is state x(t), the system will reach t_s within a finite time and then reach the equilibrium state, where:

$$t_s = \frac{n}{\beta(n-m)} |x(0)|^{(n-m)/n} \tag{10}$$

$\beta x^{m/n}$ is introduced in formula (8) so that the farther the equilibrium state is, the faster the convergence speed will be, and it will stay after reaching the equilibrium point [15]. For the first-order system finite-time control design of the mobile robot system, the control law of ω is selected as:

$$\omega = \omega_d + \beta_1 \theta_e^{\frac{m_1}{n_1}} \tag{11}$$

It can be obtained that $\theta_e = 0$ is reached within a finite time t_1, where:

$$t_1 = \frac{n_1}{\beta(n_1-m_1)} |x(0)|^{(n_1-m_1)/n_1} \tag{12}$$

It can be seen from formula (4) that as long as $t > t_1$, $\theta_e = 0$ therefore, the direction error of the robot is 0. At this time, as long as the control of the other two states is considered, formula (4) becomes:

$$\begin{cases} \dot{x}_e = y_e \omega_d + v_d - v \\ \dot{y}_e = -x_e \omega_d \end{cases} \tag{13}$$

For the mobile robot system (13), design a sliding mode controller to design the control law of v. First, design the switching function as follows:

$$s = x_e - y_e \tag{14}$$

When the equilibrium point is reached, $s \to 0$, so that x_e converges to y_e, and the error approaches 0. When $x_e = y_e$, its stability is proved by the Lyapunov function:

$$V_y = \frac{1}{2} y_e^2 \tag{15}$$

So: $\dot{V}_y = y_e \dot{y}_e = y_e(-x_e \omega_d) = -x_e^2 \omega_d \leq 0$, if and only if $x_e = y_e = 0$, $\dot{V}_y = 0$.

The exponential reaching law converges faster than the constant velocity reaching law, the general reaching law and the power reaching law, so the exponential reaching law is taken, which is expressed as:

$$\dot{s} = -\varepsilon \operatorname{sgn}(s) - ks, \, \varepsilon, k > 0 \tag{16}$$

In order to weaken chattering, the continuous function $\frac{s}{|s|+\sigma}$ is used instead of the sign function, so that express formula (16) as:

$$\dot{s} = -\varepsilon \frac{s}{|s| + \sigma} - ks, \, \varepsilon, k > 0 \tag{17}$$

From formula (13) to formula (16), we can get:

$$\dot{s} = -\varepsilon \frac{s}{|s| + \sigma} - ks = \dot{x}_e - \dot{y}_e = y_e \omega_d + v_d - v + x_e \omega_d \tag{18}$$

After sorting out, the control law of v can be obtained as:

$$V = y_e \omega_d + v_d + x_e \omega_d + \varepsilon \frac{s}{|s| + \sigma} + ks \tag{19}$$

3.2   Fuzzy Sliding Mode Controller Design

The selection of sliding mode control parameters is often difficult due to both chattering and approach time issues. In order to make the mobile robot control system have better dynamic performance and static performance, this paper uses fuzzy control to modify the input parameters, so that achieve the purpose of optimization.

The error and error rate of change are taken as the two inputs of the fuzzy controller, and the output is the parameter k in the control law of the velocity V, and the domain of input and output is taken as $[-6, 6]$. Set the position error $e_t = [e_x, e_y, e_\theta]$, the error rate of change $e_c = e_t - e_{t-1}$, The fuzzy language values are selected as: {negative big, negative medium, negative small, zero, positive small, positive medium, positive big}, recorded as {NB, NM,NS,ZO,PS,PM,PB}. According to the fuzzy set, the following membership functions are designed:

$$\mu_A(e) = \exp\left\{-\frac{(e-a)^2}{b^2}\right\} \tag{20}$$

In the formula: a and b are the constants of the membership function. The form of fuzzy control rules adopts: If x1 is A1 AND x2 is A2 Then y is C1, let $x_1 = e_t, x_2 = e_c$, $y = k$, then get a fuzzy control table, as shown in Table 1,

Table 1. Fuzzy control rule table

$e_t k e_c$	NB	NM	NS	ZO	PS	PM	PB
NB	PB	PB	PM	PM	PS	PS	ZO
NM	PB	PM	PM	PS	PS	ZO	NS
NS	PB	PM	PM	PS	ZO	NS	NS
ZO	PM	PS	PS	ZO	NS	NS	NM
PS	PS	PS	ZO	NS	NS	NM	NM
PM	PS	ZO	NS	NM	NM	NM	NB
PB	ZO	NS	NM	NM	NM	NB	NB

The weighted average decision method is used for defuzzification to obtain the precise output of the control object: $k = \frac{\sum_{i=1}^{n}\mu(k_i)k_i}{\sum_{i=1}^{n}\mu(k_i)}$ Substituting the speed control law of formula (19) into the fuzzy sliding mode control law:

$$V = y_e\omega_d + v_d + x_e\omega_d + \varepsilon\frac{s}{|s|+\sigma} + \frac{\sum_{i=1}^{n}\mu(k_i)k_i}{\sum_{i=1}^{n}\mu(k_i)}s \tag{21}$$

4 Simulation Research

Experiment with simulations on MATLAB. Select representative straight and arc reference trajectories for simulation analysis. The reference path consists of three parts. The first part is a straight line from $(-20, 4)$ to $(0, 4)$, and the second part is a right semicircle with a center at $(0, 0)$ and a radius of 4. The last part is straight line form $(0, -4)$ to $(-20, -4)$. We designed three parts with a time of 20–10–20 s. Select the velocity $v = 1.0$ m/s.

Select $\delta = 0.02, k = 2.0, \beta = 6.0, n_1 = 11.0, m_1 = 9.0$, select control law formula (11) and formula (21). In order to highlight the advantages of the control law used in this paper, $c = 5.0, \delta = 0.02, \varepsilon = 15.0, k = 20.0$ are taken. Under the same conditions, the control law formula (6) is adopted, and the simulation results are shown in Fig. 2, Fig. 3 and Fig. 4 combined with the data in Table 2.

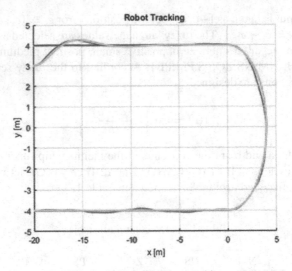

Fig. 2. Robot trajectory tracking

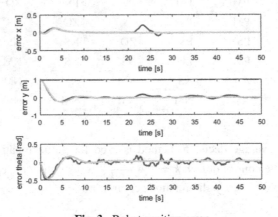

Fig. 3. Robot position error

Figure 3 shows the robot position error curve. During the robot movement, the controller gain is adjusted by fuzzy logic, which effectively reduces the system jitter and enhances the system stability, and the effect is more superior in the turning place.

Figure 4 shows the robot speed tracking curve. During the robot motion, the linear velocity remains relatively stable during the motion, and the optimized controller is smoother when reaching the critical point, and the system is more stable, and the angular velocity basically does not change.

In Table 2, v_t denotes the time when linear velocity reaches stability, ω_t denotes the time when angular velocity reaches stability, x_e, y_e, θ_e denotes the time when lateral and longitudinal errors and heading angular errors reach stable error intervals, and x_{eq}, y_{eq}, θ_{eq} denotes the error interval at which the errors stabilize.

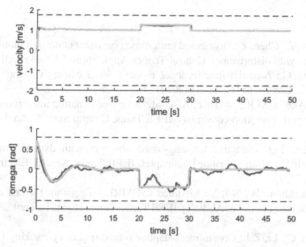

Fig. 4. Robot speed tracking

Table 2. Comparison of SMC and FSMC tracking results

	$v_t(s)$	$\omega_t(s)$	$x_e(m)$	$x_{eq}(m)$	$y_e(m)$	$y_{eq}(m)$	$\theta_e(rad)$	$\theta_{eq}(rad)$
SMC	3.9	15.6	0.1688	[−0.062,0.025]	1	[−0.032,0.158]	−0.496	[−0.125,0.083]
FSMC	2.2	11.4	0.1685	[−0.003,0.003]	1	[−0.006,0.077]	−0.421	[−0.034,0.011]

According to Fig. 2, Fig. 3 and Fig. 4, and the experimental data in Table 2, it can be seen that the actual tracking trajectory can adjust the direction in a limited time and reunite with the desired trajectory to maintain stable operation. The control law designed in this paper has the following advantages compared with the previous control law: (1) The tracking accuracy has been improved, and the error in the tracking process has been significantly reduced, especially in the process of convergence of heading angle and direction error, (2) The transformation of angular velocity and linear velocity is obviously smooth during the tracking process, so the shaking of the robot system is greatly reduced, (3) The algorithm is simple to implement.

5 Conclusion

In this paper, the fuzzy sliding mode control is applied to the trajectory tracking problem of omnidirectional mobile robot, using sliding mode variable structure control to ensure the stability of the system, using exponential convergence law to make the sliding mode with good dynamics; adding fuzzy rules and adjusting the fuzzy motion segment characteristics by the fuzzy controller to weaken the jitter; through continuous simulation experiments, it is verified that the designed controller has good robustness and better tracking accuracy, and the system is more stable; it is of great practical significance to study the trajectory tracking problem of omnidirectional mobile robot at the present stage and to find new solutions.

References

1. Jiang, Y., Liu, Z., Chen, Z.: Distributed finite-time consistent control of multi-nonholonomic mobile robots with disturbances. Control Theory Appl. **36**(5), 737–745 (2019)
2. Song, G., Tao, G.: A partial-state feedback model reference adaptive control scheme. IEEE Trans. Autom. Control **65**(1), 44–57 (2020)
3. Valasek, J., Akella, M.R., Siddarth, A., et al.: Adaptive dynamic inversion control of linear plants with control position constraints. IEEE Trans. Control Syst. Technol. **20**(4), 918–933 (2012)
4. Lai, Y.C., Le, T.Q.: Adaptive learning-based observer with dynamic inversion for the autonomous flight of an unmanned helicopter. IEEE Trans. Aerosp. Electron. Syst. **57**(3), 1803–1814 (2021)
5. Shen, C., Buckham, B., Shi, Y.: Modified C/GMRES algorithm for fast nonlinear model predictive tracking control of AUVs[J]. IEEE Trans. Control Syst. Technol. **25**(5), 1896–1904 (2017)
6. Yue, M., An, C., Li, Z.J.: Constrained adaptive robust trajectory tracking for WIP vehicles using model predictive control and extended state observer. IEEE Trans. Syst. Man Cybern. Syst. **48**(5), 733–742 (2018)
7. Gao, Z., Dai, X., Zheng, Z.: Optimal trajectory planning for mobile robot energy consumption based on motion control and frequency domain analysis. Acta Automatica Sinica **46**(5), 934–945 (2020)
8. Jeong, S., Chwa, D.: Sliding mode disturbance observer based robust tracking control for omnidirectional mobile robots with kinematic and dynamic uncertainties. IEEE/ASME Trans. Mechatron. **26**(2), 741–752 (2021)
9. Liu, J.: MATLAB Simulation of Sliding Mode Variable Structure Control. Tsinghua University Press, Beijing (2005)
10. Liu, J., He, Y.: Fuzzy global sliding mode control based on genetic algorithm and its application for flight simulator servo system. Chin. J. Mech. Eng. **20**(3), 13–17 (2007)
11. Ha Thi, K.D., Nguyen, M.C., Vo, H.T., et al.: Trajectory tracking control for four-wheeled omnidirectional mobile robot using Backstepping technique aggregated with sliding mode control. In: 2019 First International Symposium on Instrumentation, Control, Artificial Intelligence, and Robotics (ICA-SYMP 2019), pp. 131–134 (2019)
12. Peng, S., Shi, W.: Adaptive fuzzy output feedback control of a nonholonomic wheeled mobile robot. IEEE Access 43414–43424 (2018)
13. Yang, G., Liu, H., Liu, B.: Fuzzy sliding mode trajectory tracking control of four-wheeled mobile robot. Agric. Equip. Veh. Eng. **59**(9), 38–42 (2021)
14. Liu, Y.: Research on sliding mode trajectory tracking control of wheeled mobile robots. J. Jinling Inst. Sci. Technol. **25**(3), 35–38 (2009)
15. Li, S., Tian, Y.: Trajectory tracking control of moving car. Control Decis. Making **15**(5), 626–628 (2000)

Image Classification Method Base on Contrastive Learning

Junye Cao, Dongxiang Chi[✉], and Jingxuan Han

Shanghai DianJi University, Shanghai 201306, China
{216003010116,226003010318}@st.sdju.edu.cn, chidx@sdju.edu.cn

Abstract. This paper presents an innovative method for self-supervised learning in image classification, leveraging the contrastive learning paradigm. The proposed framework incorporates the SEResNet50 backbone and employs the contrastive loss function to facilitate the learning of discriminative features. Additionally, a local self-attention mechanism is introduced in the classification head following transfer learning. This work contributes in two main aspects: firstly, the proposed framework demonstrates an approximate 3% improvement in classification accuracy compared to the baseline approach; secondly, it significantly reduces training time and enhances convergence rate. Experimental evaluations on the STL-10 dataset validate the superior performance of the proposed framework over the baseline approach. Moreover, the local self-attention mechanism proves to be effective in enhancing the discriminative power of the learned features. In conclusion, this paper introduces a novel framework for self-supervised learning, which combines the SEResNet50 backbone, contrastive loss function, and local self-attention mechanism. The proposed approach achieves exceptional performance and reduces training time, thus exhibiting great promise for image classification tasks.

Keywords: Self-supervised Learning · Contrastive Learning · Local Self-attention · Image Classification

1 Introduction

Self-supervised learning has emerged as a promising approach for learning representations from unlabeled data. Unlike supervised learning, which relies on labeled data, self-supervised learning exploits the inherent structure of data to learn meaningful representations. One of the most popular self-supervised learning paradigms is contrastive learning, which learns representations by maximizing the similarity between positive pairs and minimizing the similarity between negative pairs.

In this paper, we propose a novel method for self-supervised learning that leverages the contrastive learning paradigm for image classification. Specifically, we use the SEResNet50 backbone and the contrastive loss function to learn discriminative features from unlabeled data. To further enhance the discriminative power of the learned features,

J. Li et al. (Eds.): 6GN 2023, LNICST 554, pp. 217–223, 2024.
https://doi.org/10.1007/978-3-031-53404-1_19

we introduce a local self-attention mechanism in the classification head after transfer learning.

Our proposed framework addresses the limitations of existing self-supervised image classification approaches, such as long training times and suboptimal performance. By utilizing the SEResNet50 backbone and contrastive loss function, our proposed framework learns more discriminative features from unlabeled data. Additionally, the local self-attention mechanism improves the discriminative power of the learned features and reduces the training time.

To evaluate our proposed framework, we conducted extensive experiments on the STL-10 dataset. Our experimental results demonstrate the superiority of our proposed method over state-of-the-art self-supervised learning approaches. Our proposed framework achieves an improvement of approximately 3% in classification accuracy compared to the baseline approach, while significantly reducing the training time and improving the convergence rate.

In summary, we present a novel framework for self-supervised learning that combines the SEResNet50 backbone, contrastive loss function, and a local self-attention mechanism. Our proposed method achieves superior performance and reduces the training time, making it a promising approach for image classification tasks.

2   Related Work

Contrastive learning can be classified into three categories based on the objects of comparison: context-level, instance-level, and cluster-level. Numerous studies have been conducted in this field. Context-level methods typically combine global and local feature information to train neural networks. DIM [1] and AMDIM [2] are representative context-level methods. CMC [3], proposed by Tian, Krishnan, and Isola et al., applies multiple data enhancements to image data to obtain multiple views and maximize the information shared by these views. Instance-level methods extend the research object to images and image pairs. When negative samples are used, samples are collected through external data structures or other images in the same batch. Representative instance-level methods include InstDisc [4], SimCLR [5], and Moco [6]. Instance-level methods without negative samples compare different crops of the same image, and employ special strategies to prevent training shortcuts or crashes. Currently, BYOL [7] and SimSiam [8] are popular in this category. Cluster-level methods calculate prototypes by clustering, and then compare the prototypes with samples to calculate the loss. Representative cluster-level methods include PCL [9] and Swav [10].

Based on the above research, this paper makes three contributions:

- I utilized the ResNet50 structure and incorporated the channel attention mechanism SE module into all residual modules to calculate the attention of each channel and emphasize the importance of color.
- To optimize the training process, we used the Contrastive loss instead of InfoNCE, which requires less computation.
- After transfer learning, we added a classification head at the end of the network with a local self-attention mechanism to attend to the spatial information of the final feature map.

3   Method

3.1   Algorithm Flow

To enhance the image before encoding it in the neural network, we randomly altered the image to create two views. Following encoding, the projector head transformed the feature into a vector space, where the contrastive loss was calculated. After completing the pre-training phase, we removed the projector head of the model and froze the other parameters. Subsequently, we added a classification head at the end of the model to perform transfer learning. An overview of the training framework is shown in Fig. 1.

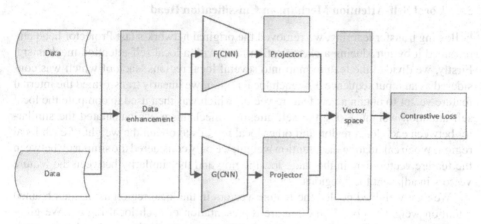

Fig. 1. Overview of the training framework

3.2   SE Module

The SE module is a widely used channel attention mechanism that boosts the expressive power of a neural network by calculating weights for each channel through a "Squeeze-and-Excitation" operation. The SE module first conducts global average pooling on each channel to obtain a global feature vector. It then learns the weights for each channel via two fully connected layers and applies them to the feature map to produce an enhanced feature map.

3.3   Contrastive Loss

Contrastive Loss is significantly faster to compute than InfoNCE Loss because the latter requires calculating the mutual information of the entire batch, while the former only needs to measure the distance between each sample and other samples. InfoNCE Loss typically requires a large batch size to achieve good performance due to its need to compute the mutual information of the whole batch. By contrast, using the same number of samples, Contrastive Loss can utilize the data more effectively and enable training with a smaller batch size. The function of Contrastive Loss is:

$$L = \frac{1}{2N} \sum_{n=1}^{N} yd^2 + (1 - y)max(margin - d, 0)^2 \tag{1}$$

d represents the Euclidean distance between the two samples. y indicates whether the two samples match, $y = 1$ indicates that they match, and $y = 0$ indicates that they do not match. *Margin* is a threshold that can be set and margin $\in (0,1]$.

3.4 Local Self-Attention Mechanism Classification Head

Following transfer learning, we removed the original network's last Projector head and retrained it by introducing a classification head with a local self-attention mechanism. Firstly, we divided the feature map into several local regions, each of which was considered as an input sequence. For each local region, we linearly transformed the internal feature vector to obtain a new feature vector, which was then used to compute the local self-attention weight. Using the self-attention mechanism, we calculated the similarity between each local region and other local regions to obtain the weight of each local region. When calculating the attention weight, we only considered the similarity between the feature vectors within the same local region and the similarity between the feature vectors in adjacent local regions.

We then weighted pooled the feature vectors in the local region using the obtained attention weights to obtain the feature representation of each local region. We globally pooled the feature representations of all local regions to obtain a fixed-size vector. We mapped the global pooled vectors to the class probability space to obtain the final classification result.

4 Experiments

4.1 Experimental Setting

For our experiment, we selected the STL-10 dataset, which contains 10 categories, 100,000 unlabeled images, 5,000 images in the training set, and 8,000 images in the test set. We pretrained our model on the unlabeled data and performed transfer learning, training the classification head on the supervised learning task using the training set. These parameter settings are shown in Table 1.

4.2 Experimental Result

According to the above parameter Settings, the following experimental results are obtained (Fig. 2).

As shown in the convergence effect diagram of the pre-training process, Contrastive Loss outperforms InfoNCE Loss in terms of convergence, especially when the batch size is small. Additionally, the pre-training process utilizing SEResNet50 as the backbone

Table 1. Experimental Setting.

Parameter	Value
projector output	128
classification output	10
pretrain epoch	800
learning rate	0.0003

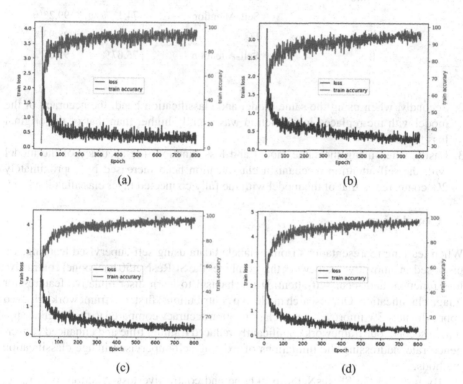

Fig. 2. Convergence of pretraining. (a) ResNet50 + InfoNCE. (b) ResNet50 + Contrastive Loss. (c) SEResNet50 + InfoNCE. (d) SEResNet50 + Contrastive Loss.

network is smoother, with no significant oscillations. Overall, the findings demonstrate that the approach proposed in this paper is more effective for pre-training when the batch size is small.

After the pre-training, delete the projector head of the original network and add a classification head to continue the training. The experimental results are shown in Table 2.

According to the results in Table 2:

1. Firstly, when using the same loss function and classification head, the Top1 accuracy of SE-ResNet50 was 1%–3% higher than that of the original ResNet50.

Table 2. Experimental Setting.

Model	Loss	Classification head	Top1	Top5
ResNet50	InfoNCE	FC	69.04%	97.71%
		Self-Attention	72.01%	98.48%
	Contrastive Loss	FC	72.37%	98.50%
		Self-Attention	74.36%	99.03%
SEResNet50	InfoNCE	FC	72.79%	98.56%
		Self-Attention	74.17%	99.25%
	Contrastive Loss	FC	73.11%	98.93%
		Self-Attention	**75.07%**	**99.35%**

2. Secondly, when using the same model and classification head, the accuracy of the model with the replaced loss function was slightly higher than that of the original loss function.
3. Lastly, when using the same model and loss function, the accuracy of the model with the self-attention mechanism classification head increased by approximately 2% compared to that of the model with the fully connected layer classification head.

5 Conclusion

When learning representations from unlabeled data using self-supervised learning, we proposed an innovative framework that combines the SEResNet50 backbone, contrastive loss function, and local self-attention mechanism to learn discriminative features for image classification. Our research made two contributions: first, our framework achieved approximately 3% improvement in classification accuracy compared to the baseline approach; second, our framework significantly reduced training time and enhanced convergence rate, addressing the limitations of existing self-supervised image classification methods.

By utilizing the SEResNet50 backbone and contrastive loss function, our framework learned more discriminative features from unlabeled data. Moreover, the local self-attention mechanism effectively enhanced the discriminative power of the learned features and reduced training time.

In conclusion, our study proposed a novel self-supervised learning framework that combines the SEResNet50 backbone, contrastive loss function, and local self-attention mechanism. Our proposed method exhibited outstanding performance and reduced training time, providing great promise for image classification tasks. Our research contributes to the field of self-supervised learning by proposing a novel approach that achieves superior performance and addresses the limitations of existing methods. We believe that our study provides useful insights for further development of self-supervised learning.

References

1. Hjelm, R.D., et al.: Learning deep representations by mutual information estimation and maximization. arXiv preprint arXiv:1808.06670 (2018)
2. Bachman, P., Hjelm, R.D., Buchwalter, W.: Learning representations by maximizing mutual information across views. In: NIPS, pp. 15509–15519 (2019)
3. Tian, Y., Krishnan, D., Isola, P.: Contrastive multiview coding. arXiv preprint arXiv:1906. 05849 (2019)
4. Wu, Z., Xiong, Y., Yu, S.X., Lin, D.: Unsupervised feature learning via non-parametric instance discrimination. In: CVPR, pp. 3733–3742 (2018)
5. Chen, T., Kornblith, S., Norouzi, M., Hinton, G.: A simple framework for contrastive learning of visual representations. arXiv preprint arXiv:2002.05709 (2020)
6. He, K., Fan, H., Wu, Y., Xie, S., Girshick, R.: Momentum contrast for unsupervised visual representation learning. arXiv preprint arXiv:1911.05722 (2019)
7. Grill, J.-B., et al.: Bootstrap your own latent: a new approach to self-supervised learning. arXiv preprint arXiv:2006.07733 (2020)
8. Chen, X., He, K.: Exploring simple Siamese representation learning. arXiv preprint arXiv: 2011.10566 (2020)
9. Li, J., Zhou, P., Xiong, C., Socher, R., Hoi, S.C.: Prototypical contrastive learning of unsupervised representations. arXiv preprint arXiv:2005.04966 (2020)
10. Caron, M., Misra, I., Mairal, J., Goyal, P., Bojanowski, P., Joulin, A.: Unsupervised learning of visual features by contrasting cluster assignments. arXiv preprint arXiv:2006.09882 (2020)

Research on RFID Indoor Localization Algorithm Based on Virtual Tags and Fusion of LANDMARC and Kalman Filter

Wu Jiangbo[✉], Liu Wenjun, and Liu Hong

Shanghai Dianji University, Shanghai, China
wujiangbi0509@163.com

Abstract. With the increase in demand for indoor positioning accuracy, the traditional LANDMARC algorithm introduces numerous tags leading to interferences between them. To address this issue and reduce costs, a new RFID indoor positioning algorithm was proposed. This novel approach was based on the integration of the Kalman filter and LANDMARC algorithm, along with the introduction of virtual tags. The primary aim of this algorithm was to reduce deployment cost and positioning errors while achieving more precise tag motion and position change descriptions. Moreover, this algorithm utilized the signal strength model of LAND-MARC and the estimation results of the Kalman filter to infer and correct the target position with nuance. Simulation experiments show that the algorithm produces reliable results with high positioning accuracy, robustness, and adaptability.

Keywords: Indoor Localization · RFID · Kalman Filter · LANDMARC

1 Introduction

Due to advancements in satellite positioning technology, both global positioning systems (GPS) and the Beidou satellite navigation system are widely used for outdoor positioning [1]. However, indoor or obstructed areas may result in poor satellite signal reception, leading to significant positioning errors. In order to cope with this issue, indoor positioning and tracking technologies have emerged, including ultra-wideband (UWB) and radio frequency identification (RFID). UWB systems possess the ability to overcome radio signal multipath distortion via filtering, providing precise positioning results as compared to traditional RFID systems. Nonetheless, despite its superiority over RFID, the high cost of constructing and maintaining UWB systems impedes their popularity [2]. Therefore, indoor positioning methods primarily use the Landmarc algorithm, which employs a dynamic active RFID calibration approach for location identification [3]. The LANDMARC algorithm has gained significant recognition in the realm of indoor positioning for its accessibility and simplicity in locating tags.

Michigan State University proposed LANDMARC [4], a mature indoor positioning system in 2003 that improves reader accuracy using reference tags and a weighted

J. Li et al. (Eds.): 6GN 2023, LNICST 554, pp. 224–235, 2024.
https://doi.org/10.1007/978-3-031-53404-1_20

algorithm system to determine the coordinates of the tag through a comparison of values to those of nearby reference tags. In comparison to traditional RFID positioning methods, the primary advantage of the LANDMARC system is its use of cheaper additional RFID tags, avoiding the usage of an excessive number of costly RFID readers, while maintaining high precision and reliability. The system can quickly adjust to a dynamic environment and mitigate the influence of environmental conditions, including atmospheric conditions and the behavior of particles in unobstructed environments, resulting in greater adaptability and robustness. LANDMARC is widely recognized in the indoor positioning field for its affordability, strong environmental adaptability, and reliable precision [4].

The Kalman filter algorithm is a state-based filtering technique widely used in positioning and navigation [1]. This algorithm achieves accurate estimation and prediction of the target state by fusing system models with measurement data. The Kalman filter algorithm calculates the optimal state estimate value by repetitively updating based on current observation values and system dynamic models [1]. By integrating with the indoor positioning technique, the Kalman filter algorithm can forecast target position and trajectory information by resolving uncertainties and reducing noise values to enhance the precision and robustness of the positioning system.

The practical implementation of the RFID indoor positioning system faces several challenges such as multipath effects, complex indoor environment, and multi-user interference, which may affect the practicality, robustness, and accuracy of the positioning system [6]. There are numerous works that discuss RFID indoor positioning, including several references [5–10]. Reference [6] proposed a positioning algorithm based on robust vector regression, representing good performance in resolving signal changes and outliers. However, the algorithm requires precise model parameter adjustment to optimize tracking accuracy. In contrast, reference [7] implemented a method that combines RFID technology and differential received signal strength indicator (RSSI) data for tracking or positioning. It is notable for its low cost, high reliability, and superior accuracy, but it is highly sensitive to signal interference and multipath effects. To enhance the positioning system's accuracy during the use of irregular deployment of reference tags, reference [8] posed a quadratic weighted positioning calculation method that partitions the reference area into geometrically segmented partitions and then performs quadratic weighted averaging. Reference [9] on the other hand, proposed a proximal policy optimization algorithm that utilizes shear probability ratio and stochastic gradient updates for positioning, improving accuracy and speed in large-scale positioning. Reference [5] proposed a two-stage method based on fuzzy and evolutionary clustering for interpreting RFID tracking data. This method applies fuzzy clustering to discover clusters and establishes a clustering tree through the utilization of an evolutionary clustering algorithm that leverages the principles of particle swarm optimization. Finally, reference [10] proposed a method that leverages upgraded networks and deep learning to improve positioning accuracy and precision.

While the traditional LANDMARC algorithm improves positional accuracy, it incurs high deployment costs by introducing a large number of reference tags, potentially increasing signal interference between tags and negatively impacting positioning accuracy. Therefore, this paper proposes a solution that relies on virtual tags and the fusion of Kalman filtering and the LANDMARC algorithm to reduce the density and number of reference tags, filter out noise, improve the accuracy of positioning, reduce deployment costs, and enhance system robustness. It aims to lower the cost of RFID indoor positioning while also improving accuracy.

2 Related Work

2.1 Radio Frequency Identification (RFID) Technology

As electronic signals propagate from the reader to electronic tags, they encounter a loss of power during transmission. This signal attenuation follows a logarithmic distance path loss model, which is represented in formula (1):

$$PL(d) = PL(d_0) + 10nlg\left(\frac{d}{d_0}\right) + X_\sigma \tag{1}$$

where, $PL(d_0)$ denotes the path loss of the tag signal at a distance d_0; X_σ denotes a Gaussian distributed random variable that has an average value of 0; and n represents the path loss factor. The path loss factor is determined by the indoor environment and obstacles, and typically ranges from 1.8 to 4.2. Once the RFID tag signal has undergone path loss attenuation, the correlation between the distance d at the reader and the received signal strength $Pr(d)$ can be expressed using formula (2):

$$Pr(d) = Pr(d_0) - 10nlg\left(\frac{d}{d_0}\right) - X_\sigma \tag{2}$$

Here, the received signal strength from the tag at the reader when they are at a distance of d_0 is denoted by $Pr(d_0)$. For the sake of computational convenience, $d_0 = 1$ is typically set to 1.

2.2 LANDMARC Algorithm

The LANDMARC algorithm employs a weighted strategy based on RSSI values of reference tags in order to estimate the position of a target tag. By comparing the RSSI values between the reference tags and the target tag, the algorithm achieves precise localization. When a tag transmits a signal, the reader can utilize the RSSI value to ascertain its position. This is particularly relevant when the target tag and a reference tag are situated in close proximity to each other, their signal values at the reader will be similar. By knowing the coordinate information of the reference tags, reference tags with signal values similar to the tag being located can be identified, thereby eliminating interference from nearby environmental factors on the tag signal. Finally, a weighted algorithm is employed to work out the coordinates of the target tag, thereby avoiding excessive signal interference and deployment costs that may arise from using too many reference tags.

Consider an RFID indoor positioning system that consists of x readers, y reference tags, and z tags to be located. The RSSI matrices for the reference tags and the tags to be located at each reader are denoted as $E_j = [E_{1j}, E_{2j}, ..., E_{yj}]$ and $\theta_j = [\theta_{1j}, \theta_{2j}, ..., \theta_{yj}]$, respectively. As each reader can capture signals from y reference tags, the total number of received signals across the x readers amounts to $x * y$. The matrix E represents the RSSI values corresponding to the reference tags:

$$E = \begin{Bmatrix} E_{11} & E_{12} & \cdots & E_{1x} \\ E_{21} & E_{22} & \cdots & E_{2x} \\ \vdots & \vdots & \ddots & \vdots \\ E_{y1} & E_{y2} & \cdots & E_{yx} \end{Bmatrix} \tag{3}$$

Similarly, the signal value matrix for the target tag can also be acquired, denoted as θ:

$$\theta = \begin{Bmatrix} \theta_{11} & \theta_{12} & \cdots & \theta_{1x} \\ \theta_{21} & \theta_{22} & \cdots & \theta_{2x} \\ \vdots & \vdots & \ddots & \vdots \\ \theta_{z1} & \theta_{z2} & \cdots & \theta_{zx} \end{Bmatrix} \tag{4}$$

In the matrices mentioned above, E_{ij} represents the signal value of the i-th reference tag on the j-th reader, whereas θ_{ij} corresponds to the signal value of the i-th tag to be located on the j-th reader. In real-world scenarios, signal strength is typically measured using numerical values. To compute the Euclidean distance between the i-th reference tag and j-th tag to be located, we employ the Euclidean distance formula (5):

$$D_{ij} = \sqrt{\sum_{k=1}^{m} (\theta_{kj} - E_{ki})^2} \tag{5}$$

Here, D_{ij} represents the distance between the i-th reference tag and the j-th tag to be located. Smaller values of D_{ij} indicate closer proximity between them, and larger values suggest a greater distance. By performing correlation operations, obtained the correlation matrix for the reference tags and tags to be located (6)

The variable D_{ij} denotes the distance between the i-th reference tag and the j-th tag to be located. Smaller values of D_{ij} indicate a closer proximity, whereas larger values imply a greater distance. Applying correlation operations yields the correlation matrix for both the reference tags and target tags (6):

$$D = \begin{Bmatrix} D_{11} & D_{12} & \cdots & D_{1x} \\ D_{21} & D_{22} & \cdots & D_{2x} \\ \vdots & \vdots & \ddots & \vdots \\ D_{y1} & D_{y2} & \cdots & D_{yx} \end{Bmatrix} \tag{6}$$

Each row of the matrix reflects the relationship between a certain tag to be located and the entire set of reference tags. By applying K Nearest Neighbor algorithm and weighting the adopted reference tags, the coordinates of the target tag's position can be determined.

$$(x, y) = \sum_{i=1}^{k} w_i(x_i, y_i) \tag{7}$$

$$w_i = \frac{1/D_i}{\sum_{i=1}^{k} 1/D_i} \tag{8}$$

The weight of the i-th neighboring reference tag is denoted by w_i, and the coordinates of the reference tag are (x_i, y_i). (x, y) represents the coordinates of the tag that needs to be located.

The weight assigned to the i-th neighboring reference tag is represented by w_i, and its coordinates are (x_i, y_i). The coordinates (x, y) correspond to the tag to be located. The estimation error, which represents the distance between the estimated and actual locations, is given by formula (9).

$$e = \sqrt{(x - x_r)^2 + (y - y_r)^2} \tag{9}$$

(x_r, y_r) represent the true coordinates of the target tag.

3 Methodology

3.1 Optimize the Layout of Virtual Tags

The LANDMARC algorithm's primary concept is to enhance positioning accuracy by utilizing high-density reference tag layouts. Nevertheless, these layouts present issues such as increased hardware costs, frequent multi-tag collisions, complex computation, and intensified multipath effect interferences which deteriorate the system's stability and anti-interference abilities. For this reason, this paper proposes a method to reduce reference tag deployment density, thereby diminishing tag collisions and multipath effects while improving the system's stability. This approach increases flexibility in tag deployment, reduces deployment costs and complexity. However, it should be emphasized that reducing the deployment density can adversely affect the accuracy of positioning.

Currently, there are two main methods for inserting virtual reference tags: grid-based and slicing insertions. However, excessive use of grid-based insertion may result in low utilization of virtual reference tags, and the accuracy of the final location result obtained through slicing insertion can be significantly influenced by the initial positioning. To address these issues, this paper presents a novel approach utilizing the circumcenter principle of a triangle. The method calculates the circumcenter of a triangle formed by three neighboring reference tags in the neighborhood and determines the position of the virtual reference tag based on this concept. This approach enhances the positioning accuracy and utilization rates through minimizing the required number of reference tags and eliminating the need for multiple positioning, thereby enhancing accuracy. By solving Eq. (10), which is based on the principle that the distance from any point on the circle to the center of the circle is equal, obtained the coordinates (x_a, y_a) of the point of insertion of the virtual reference tag. This insertion point is also known as the circumcenter a, which is the location where the virtual reference tag is inserted.

$$\begin{cases} (x_A - x_a)^2 + (y_A - y_a)^2 = R^2 \\ (x_B - x_a)^2 + (y_B - y_a)^2 = R^2 \\ (x_C - x_a)^2 + (y_C - y_a)^2 = R^2 \end{cases} \tag{10}$$

Get:

$$\begin{cases} x_a = (C_1 \cdot B_2 - C_2 \cdot B_1)/(A_1 \cdot B_2 - A_2 \cdot B_1) \\ y_a = (A_1 \cdot C_2 - A_2 \cdot C_1)/(A_1 \cdot B_2 - A_2 \cdot B_1) \end{cases} \tag{11}$$

where:

$$A_1 = 2 \cdot (x_B - x_A) \quad A_2 = 2 \cdot (x_C - x_B)$$

$$B_1 = 2 \cdot (y_B - y_A) \quad B_2 = 2 \cdot (y_C - y_B)$$

$$C_1 = x_B^2 + y_B^2 - x_A^2 - y_A^2 \quad C_2 = x_C^2 + y_C^2 - x_B^2 - y_B^2$$

where (x_A, y_A), (x_B, y_B), (x_C, y_C) are the coordinates of neighboring tags.

3.2 Optimization of Tag Signal

The localization method proposed previously is based on RSS measurements at each time point to determine the tag's position. However, the presence of noise and outliers can lead to deviations between the estimated position and the true position, resulting in an unsmooth trajectory of tag movement. To enhance positioning accuracy, this paper employs a dynamic localization or tracking method, which uses the smoothness movement trajectory of the tag and motion dynamics principles for accurate position prediction. The utilization of the Kalman filter process aims to improve tracking accuracy by effectively incorporating the current RSS measurement with the previous state. The Kalman filter can be implemented recursively by estimating the optimal state using a two-step approach. Firstly, the tag's position is estimated using a prediction model and then estimates it using a linear measurement model and the current measurement. Since the algorithm can combine the current measurement and the previous state, and has recursive properties, it can estimate the current state in real time and provide more accurate positioning. This paper employs a second-order kinematic model to capture the motion characteristics of the tag. The advantage of this model is that the speed is constant, however, the motion model of the tag incorporates an acceleration noise term. The state variable, denoted as $u(t) = \begin{bmatrix} p_x(t) \ v_x(t) \ p_y(t) \ v_y(t) \end{bmatrix}^T$, consists of position coordinates $p_x(t), p_y(t)$ and velocities $v_x(t), v_y(t)$. The motion model is described by the following equations:

$$u(k) = Fu(k-1) + w(k) \tag{12}$$

where,

$$F = \begin{bmatrix} 1 & \Delta & 0 & 0 \\ 0 & 1 & 0 & 0 \\ 0 & 0 & 1 & \Delta \\ 0 & 0 & 0 & 1 \end{bmatrix}, \quad Q = \begin{bmatrix} \frac{1}{3}\Delta^3 & \frac{1}{2}\Delta^2 & 0 & 0 \\ \frac{1}{2}\Delta^2 & \Delta & 0 & 0 \\ 0 & 0 & \frac{1}{3}\Delta^3 & \frac{1}{2}\Delta^2 \\ 0 & 0 & \frac{1}{2}\Delta^2 & \Delta \end{bmatrix} \epsilon^2 \tag{13}$$

The parameter Δ represents the sampling rate of RSS measurements and $w(k) \sim N(0, Q)$ represents white Gaussian acceleration noise. The selection of the parameter ϵ relies on estimating the variation in speed, $\Delta V \approx \sqrt{T\epsilon^2}$, over a specific time interval T. Therefore, the parameters are related to the speed of movement of the RFID system and objects.

The predictive equation employed by the Kalman filter can be expressed as:

$$u(k \mid k - 1) = F\hat{u}(k - 1 \mid k - 1) \tag{14}$$

$$P(k \mid k - 1) = FP(k - 1 \mid k - 1)F^T + Q \tag{15}$$

Here, $P(k \mid k - 1)$ denotes the error covariance of the predicted state $u(k \mid k - 1)$, derived from the preceding state estimate $\hat{u}(k \mid k - 1)$. The Kalman filter is also known as the optimal linear filter, but in RFID positioning systems, To address the nonlinear decay of signal strength in reference tags, a method utilizing second-order Lagrange interpolation was proposed to estimate the RSSI values of virtual reference tags. This approach aims to bring the RSSI values closer to the true values. The functional formula is shown in Eq. (16):

$$g\left(d_i^a\right) = \sum_{k=1}^{N} g\left(d_i^k\right) l_{k,n}(X) \tag{16}$$

where $g\left(d_i^k\right)$ is the RSSI value of neighboring reference tag k on reader i, and the calculation formula of $l_{k,n}(X)$ is shown in Eq. (17):

$$l_{k,n}(X) = \frac{\prod_{j=1,j\neq k}^{N}\left(X - d_i^j\right)}{\prod_{j=1,j\neq k}^{N}\left(d_i^k - d_i^j\right)} = \frac{\left(X - d_i^1\right) \cdots \left(X - d_i^{k-1}\right)\left(X - d_i^{k+1}\right) \cdots \left(X - d_i^N\right)}{\left(d_i^k - d_i^1\right) \cdots \left(d_i^k - d_i^{k-1}\right)\left(d_i^k - d_i^{k+1}\right) \cdots \left(d_i^k - d_i^N\right)} \tag{17}$$

where $N = x + y$; d_i^k represents the distance between reference tag k and neighboring reference tags to reader i, and X represents the distance from the inserted virtual reference tag to the reader.

The estimated RSSI value of the tag to be located, obtained through second-order Lagrange interpolation, can be utilized as a synthesized measurement for the Kalman filter. Assuming a Gaussian distribution for the noise in position estimation is a commonly adopted assumption in many positioning systems, allowing for the application of statistical methods, such as the Kalman filter, to effectively model and estimate the tag's position, a linear relationship can combine the synthesized measurement with the true state, resulting in the following measurement equation:

$$p(k) = Hu(k) + v(k) \tag{18}$$

Here, the $v(k) \sim N(0, R)$ follows independent white Gaussian distribution with zero mean and covariance R. The matrix H (19) is a selection matrix that selects the corresponding state variables for the observation model.

$$H = \begin{bmatrix} 1 & 0 & 0 & 0 \\ 0 & 0 & 1 & 0 \end{bmatrix} \tag{19}$$

The determination of the covariance matrix $\mathbf{R}$ for measurement noise involves analyzing the training samples of reference tags. In particular, the covariance matrix is computed by evaluating the covariance of training errors:

$$R = E\left((p - p_t)(p - p_t)^T\right) \tag{20}$$

In the above section, by utilizing the second-order Lagrange interpolation method described earlier, the estimated position p is obtained, while p_t represents the true position of the reference tag. The update of the current state of the tag follows the measurement model specified in Eq. (18):

$$\hat{u}(k \mid k) = u(k \mid k - 1) + K(p(k) - Hu(k \mid k - 1)) \tag{21}$$

$$K = P(k \mid k - 1)H^T \left(HP(k \mid k - 1)H^T + R\right)^{-1} \tag{22}$$

$$P(k \mid k) = (I - KH)P(k \mid k - 1) \tag{23}$$

Here, the error covariance of the current estimated state is represented by $P(k \mid k)$. By employing Eqs. (14), (15), (20), (21), and (22), the initial state vector of the tag can be recursively estimated, enabling the inference of its movement trajectory.

3.3 Algorithm Flow

The algorithm flowchart for RFID indoor positioning system is shown in Fig. 1:

After the iterative process in the flowchart is completed, a weighted localization technique is utilized to compute the coordinates of the target tag, as shown in Eq. (24):

$$(x_0, y_0) = \sum_{i=1}^{k} w_i'(x_i, y_i) \tag{24}$$

Here, the position of the tag to be located is represented by the coordinates (x_0, y_0), and the weight adjustment formula w_i' is obtained through the utilization of Eq. (25):

$$w_i' = \frac{1/D_i'^2}{\sum_{i=1}^{k} 1/D_i'^2} \tag{25}$$

Here, the distance D_i' refers to the Euclidean distance between the neighboring tag and the target tag.

To assess the accuracy and stability of the localization algorithm, the root mean square error (RMSE) is employed as a reliable performance metric. The RMSE quantifies the degree of deviation between the actual coordinates of the reference tags and the corresponding estimated coordinates obtained from the positioning algorithm. The RMSE formula is given by Eq. (26):

$$\sigma = \sqrt{\frac{1}{n}\sum_{i=1}^{n}(e_i' - e)^2} \tag{26}$$

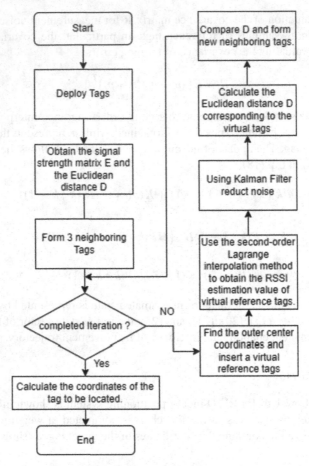

Fig. 1. System flowchart

Here, $e_i{'}$ represents the estimation error of each target tag, which is calculated using formula (27):

$$e' = \sqrt{\left((x_0 - x_p)^2 + (y_0 - y_p)^2\right)} \qquad (27)$$

Here, the true coordinates of the tag to be located are represented by (x_p, y_p), while the estimated coordinates are denoted by (x_0, y_0). The deviation between these coordinates can be quantified by calculating the average system error:

$$\bar{e} = \frac{1}{n}\sum_{i=1}^{n} e_i' \qquad (28)$$

Here, n is the total number of location estimations performed by the system.

4   Simulation and Analysis

A 20 m * 20 m indoor environment with four readers and 40 reference tags was simulated by using simulation software. Comparative experiments were conducted in this study on different algorithms, including the LANDMARC algorithm, the Quadratic Weighting with LANDMARC algorithm, the Kalman Filtering with LANDMARC algorithm, and the Kalman Filtering with LANDMARC algorithm based on virtual tags.

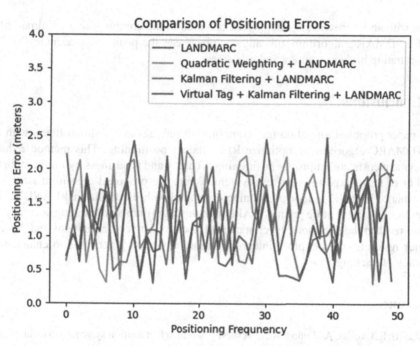

Fig. 2. Comparison chart of positioning errors

Figure 2 shows the location error of the target tags obtained from 50 simulations under different algorithms. Based on a comprehensive analysis of multiple experimental results, the combination of Kalman filtering algorithm and LANDMARC algorithm has a significant improvement in location accuracy compared to using LANDMARC algorithm alone. Even with a reduction in the number of tags used, the positioning accuracy of the Kalman filter + LANDMARC algorithm using virtual tags is comparable to that of the Kalman filter + LANDMARC algorithm using a large number of tags. Therefore, the Kalman filter + LANDMARC algorithm based on virtual tags has advantages such as high accuracy and low cost in the application of RFID indoor positioning system. Additionally, this method can reduce interference between tags, improve positioning accuracy, and increase reliability.

After analyzing the experimental results in Table 1 when the Kalman filter is used with virtual tags to complement LANDMARC algorithm, the average and root-mean-square (RMS) positioning errors are both lower than using LANDMARC algorithm

Table 1. Algorithm error comparison table.

The name of the algorithm	Average error	Root mean square error
LANDMARC	0.7783	0.8649
LANDMARC + Kalman filter	0.6515	0.7045
Virtual tags + LANDMARC + Kalman filter	0.6349	0.6916

alone, with an improvement of 19.4%. Compared to the combination of Kalman filter and LANDMARC algorithm, this approach increases the positioning accuracy by 2%, demonstrating higher accuracy and robustness.

5 Conclusions

This paper proposed a method that combines virtual tags and Kalman filters with the LANDMARC algorithm to optimize RFID indoor positioning. This method reduces deployment costs and improves positioning accuracy and robustness by minimizing the number of tags. Experimental results demonstrate the proposed algorithm's superior performance in RFID indoor positioning compared with the traditional LANDMARC algorithm, making it more practical. Although this algorithm's positioning accuracy has not yet reached the sub-meter level, it can still locate relatively large items and devices. Further research could explore integrating more advanced algorithms to enhance the accuracy of tracking.

References

1. Zafari, F., Gkelias, A., Leung, K.K.: A survey of indoor localization systems and technologies. IEEE Commun. Surv. Tutor. **21**(3), 2568–2599 (2019)
2. Dong, F., Shen, C., Zhang, J., et al.: A TOF and Kalman filtering joint algorithm for IEEE802. 15.4 a UWB Locating. In: 2016 IEEE Information Technology, Networking, Electronic and Automation Control Conference, pp. 948–951. IEEE (2016)
3. Wei, H., Wang, D.: Research on improved RFID indoor location algorithm based on LAND-MARC. In: Proceedings of the 2022 4th International Conference on Robotics, Intelligent Control and Artificial Intelligence, pp. 1294–1297 (2022)
4. Ren, J., Bao, K., Zhang, G.: LANDMARC indoor positioning algorithm based on density-based spatial clustering of applications with noise–genetic algorithm–radial basis function neural network. Int. J. Distrib. Sens. Netw. SAGE Publications Sage UK: London, England **16**(2), 1550147720907831 (2020)
5. Hu, B., Peng, H., Sun, Z.: LANDMARC localization algorithm based on weight optimization. Chin. J. Electron. **27**(6), 1291–1296 (2018)
6. Li, L., Zheng, J., Luo, W.: RFID indoor positioning algorithm based on proximal policy optimization. Comput. Sci. **2021**(48), 274–281 (2021)
7. Fazzinga, B., Flesca, S., Furfaro, F., et al.: Interpreting RFID tracking data for simultaneously moving objects: an offline sampling-based approach. Expert Syst. Appl. **152**, 113368 (2020)

8. Li, X., Zhang, Y., Marsic, I., etal.: Deep learning for RFID-based activity recognition. In: Proceedings of the 14th ACM Conference on Embedded Network Sensor Systems CD-ROM, pp. 164–175 (2016)
9. Wu, X., Deng, F., Chen, Z.: RFID 3D-landmarc localization algorithm based on quantum particle swarm optimization. Electronics **7**(2), 19 (2018). Tan, P., Tsinakwadi, T.H., Xu, Z., et al.: Sing-ant: RFID indoor positioning system using single antenna with multiple beams based on LANDMARC Algorithm. Appl. Sci. **12**(13), 6751 (2022)
10. Liu, X., Wen, M., Qin, G., et al.: LANDMARC with improved k-nearest algorithm for RFID location system. In: 2016 2nd IEEE International Conference on Computer and Communications (ICCC), pp. 2569–2572. IEEE (2016)
11. Zhou, J., Shi, J.: A comprehensive multi-factor analysis on RFID localization capability. Adv. Eng. Inform. **25**(1), 32–40 (2011)
12. Li, H., Chan, G., Wong, J.K.W., et al.: Real-time locating systems applications in construction. Autom. Constr. **63**, 37–47 (2016)
13. Jiang, T., Huang, Y., Wang, Y.: Study on improved LANDMARC node localization algorithm. In: Xie, A., Huang, X. (eds.) Advances in Electrical Engineering and Automation. AISC, vol. 139, pp. 415–421. Springer, Heidelberg (2012). https://doi.org/10.1007/978-3-642-27951-5_62
14. Li, J., Wu, Z., Wu, C., et al.: An inexact dual fast gradient-projection method for separable convex optimization with linear coupled constraints. J. Optim. Theory Appl. **168**, 153–171 (2016)
15. Omer, M., Tian, G.Y.: Indoor distance estimation for passive UHF RFID tag based on RSSI and RCS. Measurement **127**, 425–430 (2018)
16. Chai, J., Wu, C., Zhao, C., et al.: Reference tag supported RFID tracking using robust support vector regression and Kalman filter. Adv. Eng. Inform. **32**, 1–10 (2017)
17. Wu, X., Deng, F., Chen, Z.: Rfid 3D-landmarc localization algorithm based on quantum particle swarm optimization. Electronics **7**(2), 19 (2018)
18. Xu, H., Wu, M., Li, P., et al.: An RFID indoor positioning algorithm based on support vector regression. Sensors **18**(5), 1504 (2018)
19. Zhu, J., Xu, H.: Review of RFID-based indoor positioning technology. In: Barolli, L., Xhafa, F., Javaid, N., Enokido, T. (eds.) IMIS 2018. AISC, vol. 773, pp. 632–641. Springer, Cham (2019). https://doi.org/10.1007/978-3-319-93554-6_62
20. Shirehjini, A.A.N., Shirmohammadi, S.: Improving accuracy and robustness in HF-RFID-based indoor positioning with Kalman filtering and Tukey smoothing. IEEE Trans. Instrum. Meas. **69**(11), 9190–9202 (2020)
21. Li, N., Ma, H., Yang, C.: Interval Kalman filter based RFID indoor positioning. In: 2016 Chinese Control and Decision Conference (CCDC), pp. 6958–6963. IEEE (2016)
22. Xu, H., Ding, Y., Li, P., et al.: An RFID indoor positioning algorithm based on Bayesian probability and K-nearest neighbor. Sensors **17**(8), 1806 (2017)
23. Zeng, Y., Chen, X., Li, R., et al.: UHF RFID indoor positioning system with phase interference model based on double tag array. IEEE Access **7**, 76768–76778 (2019)
24. Wang, C., Shi, Z., Wu, F., et al.: An RFID indoor positioning system by using particle swarm optimization-based artificial neural network. In: 2016 International Conference on Audio, Language and Image Processing (ICALIP), pp. 738–742. IEEE (2016)

Computer Systems and Applications

Research on Information Literacy Teaching of Library MOOC Based on Constructivism

Liu Ping[(✉)]

Harbin University of Science and Technology Library, Harbin, China
liuping@hrbust.edu.cn

Abstract. The theory of constructivism is about learning, MOOC is short for massive open online courses. This paper combines the educational form of courses with the learning theory of constructivism to cultivate students' initiative and creativity, and stimulate students' active interest in learning, and promote the development of information literacy education in universities.

Keywords: theory of constructivism · MOOC · information literacy education · teaching mode · library

1 Introduction

The core part of constructivism learning theory is student-centered, which emphasizes students' active exploration and discovery of knowledge and their active construction of knowledge. Constructivism learning is actually the process of constructing knowledge by themselves.

MOOC is short for massive open online courses, and has been called a revolution that is tearing down the walls of universities. All learners only need to have classes remotely, participate discussions and complete homework to get a diploma.

The paper combines the education form of MOOC with the learning theory of constructivism. MOOC is the integration of learning resources of famous universities, and it enables college students to receive equal and fair learning opportunities. Also MOOC reflects the transformation from teacher-centered to student-centered, and reflects the change of learners from passive acceptance to active learning. The above just validates the constructivism theory, which takes students as the center, and makes students become both the discovers and the learners.

2 Analysis

2.1 Constructivism Theory, Elements and Modes

Constructivism theory is a theory of learning. Knowledge construction embodies the quantitative accumulation and the development of qualitative change. Students in universities take the initiative to learn as the subject of learners, and teachers in universities

J. Li et al. (Eds.): 6GN 2023, LNICST 554, pp. 239–244, 2024.
https://doi.org/10.1007/978-3-031-53404-1_21

play an active role in learners as instructors. Teachers use various learning resources and platforms to enhance learners' initiative to complete the construction of knowledge. Cooperative learning between learners and instructors as well as between learners and learning resources plays a key role in the construction of learners' self-knowledge. The idea of constructivism learning theory includes four aspects: First, knowledge is self-construction of the objective world based on the original experience. Knowledge cannot be passively accepted, but can only be created by learners actively. Second, the learner is the subject of learning, and the learning effect depends on the ability of self-knowledge construction. Third, the effective construction of knowledge by learners is closely related to their learning situation experience, and the instructors effectively complete the construction of knowledge with the help of various learning resource platforms. Last, collaborative learning between learners, between learners and instructors, and between learners and other teaching media plays a key role in the construction of learners' self-knowledge.

Constructivism holds that learners' self-knowledge construction generally includes four elements: situation, collaboration, dialogue and meaning.

I. Situational Elements
The real social and cultural background of learners' knowledge construction. Instructors help learners to restructure their original knowledge structure and construct new knowledge.

II. Collaborative Elements
Self-knowledge construction by learners through collaboration, including two types: learners form interest groups voluntarily for communication and cooperation; Instructors organize learners to carry out collaborative learning according to the complexity of learners' learning content.

III. Dialogue Elements
That is learners communicate with others directly or with the help of the Internet learning platform to achieve learning activities.

IV. Sharing Elements
The common knowledge themes formed by individual interest selection of learners in the network era, and the sharing process of knowledge and information is formed among learners in certain forms.

The practical application of constructivism learning theory has formed many effective learning modes, of which there are three modes with great influence: the first mode is anchored mode, in which instructors lead learners to directly feel and experience the generation of knowledge in the real environment according to their learning tasks or problems. In the top-down mode, teachers decompose tasks according to learners' learning conditions and abilities, and then provide learners with learning environment and problem-solved tools and methods. In the mode of tacit knowledge mining, learners' tacit knowledge is generally undetected. In order to turn such knowledge into explicit knowledge, learners must dig out their tacit knowledge through self-introspection and reflection, discussion, argument and discrimination with others on the Internet.

2.2 The Present Situation and Analysis of MOOC in Libraries

MOOC courses have developed to the point where they benefit from the Internet and media, addressing the needs of students. MOOC courses realize the reorganization and dissemination of knowledge and resources. At the same time, MOOC facilitates the connection between course organizers and university libraries. For example, in terms of information search and organization, and librarians can provide professional advice to course organizers and students. Therefore, under the MOOC environment, university libraries can reposition their social functions and service models.

For college students, college libraries have the responsibility and obligation to cultivate their information literacy. The MOOC is an information literacy education course that organizes students to participate in the teaching content provided by the library. The library has a rich resource professional service team and has its own unique view on the selection of course making resources. Participants in MOOC-based teaching in libraries are becoming reference and consulting providers for MOOC courses.

MOOC courses are a hybrid form of online and offline multi-way learning. Learners need a certain virtual space in the process of learning, but for some developed MOOC course platforms, learners may encounter the problem of lack of space. Therefore, libraries can provide corresponding learning space to realize the integration of library and MOOC, that is, to provide learning space. Library can not only provide virtual network space, but also give play to its own site advantages to provide learning space for learners.

3 Method

3.1 The Construction of MOOCS Teaching Mode in Library

Constructivism learning theory is to make teachers become learners' guiders and helpers, appropriately stimulate learners' interest in learning, promote learners' motivation of independent learning, and enhance learners' self-learning ability.

After the MOOC teaching course, the video will be divided according to the knowledge point. In the design of knowledge points, we pay attention to the connection between learners' original knowledge background and new knowledge, the knowledge points also run through some fashionable words on the network and the latest movies played on the Internet. For example, Huang ruhua's "Information Retrieval MOOC", an national quality course, was developed by a team. The course included movie clips to attract learners, tips to improve their quality of life, and how to use academic search engines and access professional literature for free. Another example is the national excellent course "Document Management and Information Analysis" taught by Professor Luo Shaofeng from The University of Science and Technology of China. This course describes the acquisition of network learning resources, the use of search engines, the use of personal knowledge management notes, the application of mind mapping and the quick learning of document management software.

The construction elements of the library MOOC mode should be the main bearers of the MOOC education mode in universities. At the same time, the internal power of the single element or the combined force of multiple elements can promote the development and extension of the MOOC mode in universities, so that students in universities can realize the construction of self-knowledge and the experience of MOOC learning smoothly. These elements include teaching elements, cooperation elements, sharing elements and thinking elements.

1) Teaching elements. Constructivism believes that knowledge construction is the interaction between learners and the environment. Therefore, professors should create real situations related to learning MOOC knowledge and enable learners to solve practical problems in real situations.
2) Elements of cooperation. Constructivism holds that learners can construct the meaning of the theme more accurately and comprehensively in cooperation. Libraries construct learners' self-knowledge by providing an environment, organizing learners to learn MOOC, and providing technical support to lecturers.
3) Shared elements of today's Internet age knowledge fragmentation is a common phenomenon, constructivism through knowledge sharing is an important approach to learning, learners in the learning guide and organization of the MOOC after a certain stage or learning technology, prompt summary and sharing between learners, enhance learners' knowledge construction.

3.2 Construction Mechanism and Standard

Constructivism learning theory holds that as the leader, the teacher is the helper and guide for students to construct knowledge, and should stimulate students' interest in learning and maintain their learning motivation. Constructivism has the function and significance of enhancing learners' self-learning.

For example, in the information retrieval course after MOOCs, major knowledge points and practice topics are selected for each chapter and videos are shot for 20 class hours in total, and videos for each class hour are limited to 8–10 min. The selected knowledge points should pay attention to students' memory habits and the connection between learners' knowledge experience and newly learned knowledge, so that learners can have a clear understanding of the knowledge of the video. In addition to videos, the MOOC platform provides learners with lesson plans, PPT, exercises, reference books and frequently asked questions for follow-up self-study. Learners can download resources according to their own time and the arrangement of the network environment, and they can search for accurate answers through common questions. Let learners master the progress in learning, find answers to problems, solve problems, improve the quality of their own learning, cultivate the ability of independent learning.

The standard of constructivism is that "learning is construction, construction requires innovation". In the Internet time, learning, application and innovation are indispensable in every steps. The ultimate goal is innovation, and the highest goal is also innovation.

3.3 Construction Model

For example, the teaching and research section of information retrieval course in university library can send several teachers to form a teaching team and jointly open the information retrieval MOOC course. The students should be arranged in the teaching environment reasonably, the students should be the center, and the cultivation of students' information ability and the development of students' personality should be emphasized. In order to improve the students' information literacy training of the students' information ability, information retrieval and application courses can introduce excellent MOOC with retrieval course. Part-time teachers of literature retrieval course, according to the requirements of the curriculum and individual lectures, homework assignments and reviews, directors are responsible for the master course of answering questions and discussion. This teaching mode can not only improve the quality of the lecture content, but also help students to learn after class, collect information resources, expand the content in class and meet the individual needs of students according to their ability.

I. Teachers' Teaching
As most MOOC courses are from famous universities at home and abroad, they generally provide a lot of information resources. For example, the information retrieval course of Teacher Huang will involve the use of famous database and the application of literature management software. Based on the existing database or literature management software of our school, and taking into account the professional characteristics of students, we replace and supplement the course content. In addition, the acquisition channels of foreign language literature transmission and the acquisition methods of cooperation between our library and other libraries are supplemented. The goal of information retrieval MOOC set up by our library is to cultivate students' information literacy, to focus on information retrieval and to train students on using software and tools. The goal is to equip our college students with information literacy and innovation consciousness.

II. Students' Learning
The teachers upload MOOC course network resources and related materials to a online class, and at the same time students can enter the MOOC and start the self-study process. The teaching time of MOOC courses is generally about ten minutes or half an hour, which is characterized by a large amount of information, content hopping, resource diversification, wide audience and strong randomness. After understanding the basic theoretical knowledge, students can choose more appropriate content for independent learning according to their own needs and development. If you encounter difficult problems or difficulties, you can consult the teachers through QQ, wechat and other channels. You can also participate in online and offline discussions and exchanges.

4 Conclusion

As an important theoretical basis of library information literacy teaching, constructivism puts forward new requirements for the cultivation of library information literacy in the information age, which is complementary to the information literacy education reform on campus.

Therefore, information literacy education should take students' needs as the starting point and construct an appropriate learning system according to the actual situation. On the other hand, learning resources of network media should be adopted to strengthen situational learning, optimize teaching content, cultivate students' initiative and creativity, stimulate students' active interest in learning, and promote the development of university information literacy education.

References

1. Zhang, K.: An exploration based on the reform of classroom teaching in education. J. Modern. Educ. (6), 51–56 (2019)
2. Zhang, C.: Exploring the mixed teaching mode based on MOOC. J. High. Educ. **6**, 1–8 (2021)
3. Hu, Y., Tang, G.: Research on mobile PBL teaching method based on constructivism. Chin. J. Multimedia Online Teach. (11), 30–32 (2021)
4. Mu, Z.: Research on MOOC based blended learning mode. Mod. Educ. Technol. (5), 73–80 (2021)
5. Lu, F.: A new perspective study on the reform of higher vocational English teaching. J. High. Educ. **6**, 175–177 (2020)
6. Cai, P.: A mixed teaching model for pedagogical courses based on constructive attention. Western Qual. Educ. **8**, 134–136 (2022)
7. Zhang, D.: Enlightenment of university library on MOOCO copyright. Libr. Work. **4**, 90–93 (2021)

Research on Smart Library System Based on Big Data

Liu Ping[⊠]

Harbin University of Science and Technology Library, Harbin, China
liuping@hrbust.edu.cn

Abstract. The article first explores the construction of a library's smart information service system from the viewpoint on big data pattern. The advent of the big data era will pose challenges to the practical construction of smart libraries. This article describes a big data pattern with smart information service system for library, including four levels: infrastructure level, data resource level, technology processing level and service application level. The functions and features of every levels were talked about in detailed way. How to meet the needs of users and social development in the era of big data will be the main issue of smart library research in the era of big data.

Keywords: smart library · big data · digital service · framework

1 Introduction

A smart library is a combination of technologies such as libraries, the Internet of Things, cloud computing, and intelligent devices. It is an automated library that combines with intelligent management, personalized services, efficient knowledge sharing, and intelligent sense of readers' demands. From the perspective of digital library services, smart libraries use ICT (Information Communications Technology) to realize intelligent study, personalized customized instrument, and knowledge resources for intelligent ability of library's movement. Based on different cognitive perspectives on smart libraries, they are usually built using technologies such as cloud computing, cognitive computing, sensors, high-speed wireless transmission, RF (radio frequency) technology, big data and so on. The above can achieve operating ability between readers, readers, and devices, as well as devices and devices.

At present, with the development of the era for big data and the improvement of big data technology, the whole value, density of value, usage ability, and controlling ability of big data are also showing a quick increasing direction. For the construction of library smart service system, smart service management, reader service need angle, smart service customized ability and promotion of personalized smart reading QOS (Quality of Service) promise, the big data has become an important basis to make a decision.

Based on the current research status of smart library and smart library services in the context of big data and artificial intelligence, the article constructs a library smart

J. Li et al. (Eds.): 6GN 2023, LNICST 554, pp. 245–251, 2024.
https://doi.org/10.1007/978-3-031-53404-1_22

service framework from the perspective of dual drivers of big data and artificial intelligence, including infrastructure level, data resource level, technology processing level, and service application level. The paper analyze some main abilities of every level.

2 Analysis

While ensuring the diversity of library service models, objects and content, it also makes its data massive, exponentially increasing, multi type and complex; greatly increasing the difficulty and cost of library large record collection, transmission, store, analysis, and decision. How to effectively improve the scientific and standardized level of big data collection, storage, integration, management, processing, analysis, and decision-making in libraries is also a key factor affecting the effectiveness of large data decision-making in libraries. The library big data platform is safe, reliable, and economical, also the library has the advantages of high big data processing performance and strong system compatibility. It can offer big data decision-making support sciential to readers, because of exact digital picture for individual reading activities.

Libraries have the characteristics of multiple monitoring data nodes, high management costs, complex security technologies, and business continuity in the collection, transmission, sharing, processing, analysis, and decision-making of big data. Therefore, how to track and locate, and trace big data administrators, ordinary librarians, and equipment suppliers based on the requirements of big data application and security is directly related to the effectiveness of library big data value exploration and the security of data opening process.

In the era of big data, research on smart services has also shifted from smart information services to smart data services. Although the research on the framework of library smart services may seem outdated in the context of the cloud platform architecture, multi terminal applications, and smart service presentation of the next generation library knowledge discovery system, it still has certain practical significance.

Through statistical analysis of relevant literature, it was seen that the word "big data" has emerged since 2012, and research on libraries in the environmental data in China has increased from 2014. Since from 2014 to 2020, big data-driven library smart information services have attracted widespread attention from the academic community, including smart library construction, data mining, knowledge services and university library smart services. The personalized services and related hot spots area of smart libraries have entered a steady development stage (see Fig. 1).

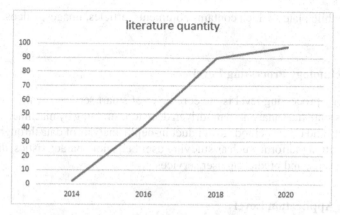

Fig. 1. Smart Library and Big Data of Literature Quantity

3 Method

The large record smart library information service system (Fig. 2) mainly includes three levels: infrastructure level, data resource level, and service application level, which relate to three constituent elements,such as information context, information, and information people in the smart information service ecosystem: (1) the infrastructure level is the material foundation and technical assistant of smart information services; (2) The data resource level provides core resources and data processing technologies for intelligent information services; (3) The service application layer is the top-level interaction port of smart information services, supplying users with big data-driven smart information services.

3.1 Infrastructure Support

The acquisition, storage, management, organization, analysis, and application of big data in the infrastructure layer rely on stable infrastructure support. The infrastructure level provides a basic support circumstance to implement smart information services driven by big data, mainly composed of hardware facilities and information technology.

3.2 The Data Resource Level

The data resource level is in charge of organizing, administering, analyzing, and mining library big data, mostly containing two aspects: library big data and big data processing technology.

1)Collection record, consisting of library numeric collections, institutional knowledge bases, scientific research management record, and so on; 2) Business management data, access control data, monitoring data, and other IoT terminal data generated from daily library operations. 3) Service record, including literature service records (search as well download records, borrowing information, consulting service records, subject service records, service evaluations. 4) User behavior record refers to the behavior record left by users on public plates, which contains comments, articles, images, videos, feedback, and so on.

3.3 The Technology Processing Level

The technology processing level is the core layer of smart services, which comprehensively utilizes statistical analysis, data-driven decision, studying by machine, natural language treating, and other methods to conduct in-depth analysis of data through the smart service integration platform, such as studying user behavior characteristics, discovering user service needs, and predicting user services.

3.4 Service Application Level

The service application level is the terminal embodiment of the entire smart service. The smart service matrix designed in this article mainly includes smart information resource services (such as information resource demand identification, information resource intelligent acquisition, information resource intelligent utilization, information resource intelligent sharing, information resource intelligent control, etc.) Smart information content services (such as smart evaluation of cultural services, knowledge graph of massive data resources, knowledge discovery, hot spot tracking in research interest fields, etc.) and smart information user services (such as user profiles, precise recommendations, service customization, etc.) are three main types to explore and accurately understand users' smart service needs, help users obtain, utilize, and share data resources, and ultimately achieve the goal of providing smart services to users.

Libraries can leverage the new connections of data to semantically label and link any data object, becoming logical and correlative knowledge characters, and constructing a large knowledge map. For instance, libraries organically associate books, papers of magazines, dependent data of papers, datasets, patents, clinical trials, colleges, meetings, writers, sponsored projects, and even downloading and reading quantities, identify, describe, and annotate the correlation between these objects, form a knowledge graph in the pattern of associated record, construct a knowledge graph based record system management plate, and achieve semantization based on knowledge maps Intelligent information services such as fine-grained intelligent retrieval, literature mining, and knowledge discovery.

In the designed library smart service framework, the sources and types of big data have been added, that is structured data, semi-structured data, and unstructured data mainly from infrastructure such as perception devices, robots, cloud platforms, and mobile internet at first; Second is the application of artificial intelligence technologies such as machine learning and natural language processing in the technical processing layer, as well as the design of personalized automatic recommendation, knowledge graph and other intelligent services based on deep learning technology support in the service application layer, as well as user profiling, hotspot tracking and other intelligent services based on big data processing and analysis applications (such as web crawlers, user experiments, log collection, etc.).

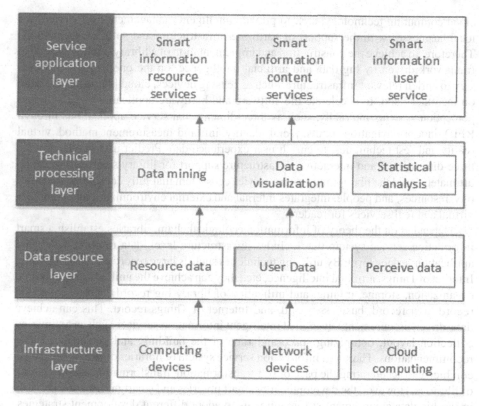

Fig. 2. Smart Library Service Framework

4 Conclusion

Smart mode has become an important tool and content in the personalized service process of users in the construction of smart libraries. With the advent of the big data era, it has become possible for libraries to measure readers' reading needs, reading behaviors, reading emotions, and reading satisfaction in detail. How can libraries effectively analyze the collected user reading behavior data and social relationship data, and accurately discover and predict the reading behavior habits, preferences, and needs of readers behind complex and scattered data, in order to build personalized smart service models, strategies, and content for users in libraries. At the same time, it is also the key to real-time adjustment and precise optimization of the library's service model and content based on changes in user needs. Therefore, libraries must establish a people-oriented service concept. Based on the individual needs of readers and the scientific analysis results of big data, we can provide readers with safe, efficient, satisfactory, and low-carbon personalized big data reading services.

At present, the smart employs of libraries only got to the level of "something intel-ligent", and the construction of infrastructure only remains on the local supplement of smart instruments for service processes, that is, relying on the Internet of Things, sensing technology of intelligent networks, intelligent perception technology, big data, cloud computing technology, etc., to provide intelligent management and services for local businesses with the support of hardware infrastructure and software technology. Therefore, to achieve the construction and implementation of a library intelligent service framework driven by big data and artificial intelligence, on the one hand, it is necessary to enrich relevant infrastructure, such as sensing devices, cloud platforms, etc., and on the other hand, to accelerate the software and hardware applications of intelligent perception systems internally, such as three-dimensional service applications through RFID, indoor navigation robots, facial identity, infrared measurement method, virtual reality and rest techniques, to enrich user experience data Predict user needs, provide three-dimensional and interactive infrastructure support for library smart services, and ultimately build the library into a smart service ecosystem that fully coordinates technology, resources, and people, integrates internal and external environments, and combines virtual and real services for readers.

Depended on the theory of information ecological chain, libraries establish a smart information service system that includes infrastructure level, data resource level, and application service level. By utilizing techniques such as large record, cloud computing, Internet of Things, artificial intelligence, etc., they can achieve the unified representation, organization, storage, mining, and utilization of library big record such as collection record, user record, business record, and Internet of Things record. This can achieve data-driven procurement Diversified intelligent information services such as optimizing collection layout, optimizing space and time of the building, and personalized user recommendations. The smart information service system of libraries driven by big data is conducive to accelerating the pace of self transformation, transformation and upgrading of libraries. However, the development of smart libraries and smart information services in the big data environment still requires us to adopt different development strategies, utilize artificial intelligence technology, build knowledge associations, and create an interactive smart service ecosystem, so that future libraries can move towards intelligent, efficient, ubiquitous, and personalized smart libraries.

Big data and artificial intelligence have deepened the connotation of library smart services and accelerated the pace of library transformation and upgrading in the new era. Smart libraries and smart services are the future development direction of libraries. However, the framework construction and service implementation of smart services in libraries still face many difficulties. We need to adopt practical and effective professional strategies to promote smart libraries towards automatic intelligence, efficient ubiquitous, and personalized refinement. The coming of the big data era and changes in user knowledge needs will pose challenges to the construction of smart libraries, but it also creates new growth points for academic research in smart libraries. Building on the research topics of how to collect, integrate, store, organize, analyze and serve data in the big data era, achieving the construction of smart libraries will undoubtedly generate more research hot spots.

References

1. Gandomi, A., Haider, M.: Beyond the hype: big data concepts, methods, and analytic. Int. J. Inf. Manage. **35**(2), 137–144 (2016)

2. Ma, J., Zhao, T., Wang, S.: Construction of functional structure model for smart libraries in universities. Inf. Sci. **8**, 56–61 (2017)
3. Shuzhen, C.: The smart service model and its implementation of smart libraries. Intell. Explor. **3**, 112–115 (2016)
4. Rousseau, R.: A view on big data and its relation to informatics. Chin. J. Libr. Inf. Sci. **3**, 12–26 (2020)
5. Tang, Y.: Survey and analysis of the integration demand for public digital cultural resources in China. Libr. Inf. Work **59**(11), 6–12 (2015)
6. Pradeep Siddappa, LibrARi. http://www.pradeepsiddappa.com/design/librar. Accessed 3 May 2022
7. Yi, C.: The ethica lgovernance dilemma and advantages of social media data research framework design. Libr. Constr. **3**, 43–45 (2019)
8. Xin, Z.: Research on the construction of government data governance system in the 5G era. Inner Mongolia Sci. Technol. Econ. **5**, 62–63 (2021)

Smoke Segmentation Method Based on Super Pixel Segmentation and Convolutional Neural Network

Wang chengkun[✉], Zhang jinqiu, Yang jiale, and Feng kaiyue

Heilongjiang University of Science and Technology, Harbin 150040, China
402686820@qq.com

Abstract. The steps of the fire disaster are from smokes to open flame. It has significant and practical meaning to use fixed camera stands to detect smoke. As the rapid development of AI in recent years the methods that using deep learning to monitor smoke pixels has owned technical foundation, But compared with the flame, the smoke has more complex pixel gray value. Segmentation results often affected by fog, water mist, clouds and other factors. On the other hand, as the deep learning is belong to supervision learning, we need to mark the smoke pixels before the model training. It is easy to mismark pixels due to man-made factors during the practical operation. So as to solve these problems mentioned, this paper regarded single frame with smoke as research object, using SegNet model of deep leaning to split the smoke pixel from the image, and then divide the pixels contained smoke into blocks by simple linear iterative clustering(SLIC).In the end, we combine the result of super pixel segmentation with SegNet model. The experimental results show that the results obtained by the above method is better than the original SegNet and the details of the smoke can be better reflected.

Keywords: Smoke Segmentation · Segnet · Slic · Smoke image

1 Introduction

Fire disaster brought great harm to people's safety of life and property, and it's necessary to take earlier warning on the beginning of the fire and have a accurate to the state of the fire. As fires often start with smoke followed by open flames. So the detection to smoke characteristic can not only effectively prevent the happening of the fire, but also brought some help to the fire extinguish when the fire break out.

The characteristics of smoke detection can not only effectively prevent the happening of the fire, but can also bring fire put out of work when the fire broke out. With the development of machine vision technology, the use of monitoring and control systems to monitor smoke can replace artificial intelligence effectively. The segmentation to image containing smoke can not only determine the location of smoke but also the development and state in time. At present, there are two main categories of smoke image segmentation: traditional method and artificial intelligence.

Chen Shouhe et al. [1], propose an early fire alarm method based on video processing. The basic idea of the proposed fire detection is to use chromaticity and disorder measurements based on the RGB (red, green, blue) model to extract fire pixels and smoke pixels.Flame pixels of the decision function are mainly composed of the strength of the R component and saturation is derived.Extraction of the fire pixels will be verified, if it is a real fire growth and chaotic dynamics, and smoke.By iteratively checking the flame growth rate, the fire alarm is given when the alarm condition is met.Walter Phillips III et al. [2], used video sequences computed with color and motion information to locate fire.Then the temporal variation on pixels is used to determine which of these pixels are actually fire pixels. Finally, an erosion operation is used to automatically remove some false fire pixels and a region growing method is used to find some missing fire pixels.

Smoke recognition methods based on deep learning have been widely used in recent years. Wang Zilong et al. [3], proposed the use of external smoke images and deep learning algorithms to explore real-time prediction of transient fire scenes, solving the problem of how to determine hidden fire information using smoke images. Abdusalomov A B et al. [4], improve the Detectron2 model by using custom data sets and 5200 images in various experimental scenarios to detect smoke at a long distance during the day and night. Saponara S [5] put forward a kind of based on convolution YOLOv 2 neural network real time video fire and smoke detection methods. Larsen A et al. [6], used deep fully convolutional neural networks of as the method of predicting fire smoke in satellite images. Khan S [7] put forward a clear based on convolution neural network (CNN) of smoke and fog environment detection and segmentation framework;Yar H et al. [8], propose a new framework based on lightweight convolutional neural network (CNN), which requires less training time and is suitable for resource-constrained devices.Zhang Lin et al. [9], proposed a classification-assisted gate recursive network (CGRNet) semantic segmentation method for smoke, which outputs smoke with two predicted probabilities from the classification module. Dubey V et al. [10], Proposed an early fire detection model and has been applied to Raspberry PI microcontroller to some degree. Avazov K et al. [11], proposed a technique that can perform early fire detection and classification to ensure maximum safety of organisms in forests.

Most of the above methods belong to supervised learning methods, and it is inevitable to label smoke images when training the model. However, the smoke image itself is very complex, and its texture and shape are very similar to that of smog, cloud and fog, so it is difficult to distinguish in detail, and errors will inevitably occur in labeling. On the other hand, in images containing smoke, there are also some cases where the smoke is very small and not easy to capture. In order to solve the above problems, this paper combines the super pixel segmentation method with the semantic segmentation in the deep learning method, in order to further improve the smoke segmentation effect.

2 Method

2.1 SegNet Smoke Image Segmentation Method

SegNet is a deep convolutional neural network for image segmentation. SegNet is improved on the basis of VGG16 network. By using the encoder-decoder structure, the input image is down-sampled (encoder) and up sampled (decoder), so as to achieve the purpose of predicting smoke in the image. Figure 1 shows the basic principles of SegNet.

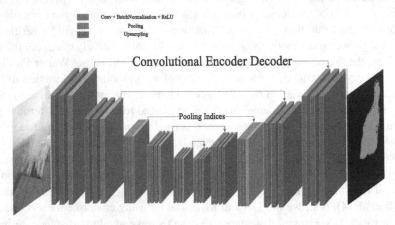

Fig. 1. Basic principle of SegNet

As shown in Fig. 1, the process of SegNet consists of input image, Encoder network, Decoder network, softmax classification, and output segmentation image. It is generally divided into two parts: the encoder responsible for input and the decoder responsible for output.

The Encoder part of SegNet is essentially a series of convolutional networks, which consists of convolutional layer, pooling layer and Batch Normalisation-layer. The volume base layer is responsible for obtaining the local features of the image, the pooling layer sub samples the image and transmits the scale-invariant features to the next layer, and BN mainly normalizes the distribution of the training image to accelerate the learning. Therefore, encoder can classify and analyze low-level local pixel values of images, so as to obtain higher-order semantic information. The Decoder in SegNet upsamples the reduced feature image, and then convolutional processing is performed on the upsampled image. The purpose is to improve the geometry of the object and make up for the loss of details caused by the pooling layer in the Encoder. The Decoder part adopts a structure that is symmetric to the Encoder part, and the up sampling operation is performed at the corresponding position of each Encoder layer and the up sampling results are convolved using the corresponding weights. These weights are copied from the corresponding Encoder convolution layer, to maintain the continuity of feature mapping.

2.2 SLIC Super Pixel Segmentation Method

Super pixel segmentation refers to the process of sub dividing a digital image into several sub-intervals. SILC in super pixel segmentation is an algorithm with simple idea and easy to implement. The method is described as follows.

The RGB image was first converted to the CIELAB color space, where L represents the brightness of the color, a represents the position between red and green, and b represents the position between yellow and blue.

Then the image is evenly divided into several small areas, assuming that the image $A(M \times N \times 3)$, And it is segmented into K super pixels of the same size and size, and the number of pixels in each super pixel is $W = (M \times N)/K$, For convenience, select the small square area of size $S \times S$ when selecting the super pixel size, and ensure M/S and N/S be an integer.

Next, select the specific location of the clustering center. In order to avoid the clustering center falling on the contour boundary with a large gradient, move the clustering center to the place with the smallest gradient inside the little square of $S \times S$.

Each pixel is then assigned a class tag related to the previously assigned seed points by searching for each pixel $2S \times 2S$ cluster center in neighborhood range, and the distance between the pixel and the cluster center is solved according to formulas (1), (2) and (3).Formula (1) reflects the distance between the cluster center and the pixel in the pixel value, where(l_j,a_j,b_j),It's the number one in the graph$j(j \in N^+ \cap [1,M \times N])$the LAB value of each pixel,(l_i,a_i,b_i)represents $2S \times 2S$ the LAB value of the i cluster center in the neighborhood range.

$$d_c = \sqrt{(l_j - l_i)^2 + (a_j - a_i)^2 + (b_j - b_i)^2} \tag{1}$$

Formula (2) reflects the spatial distance between clustering center and pixel point, where (x_i,y_i),represents the first in the image,the spatial position of $j(j \in N^+ \cap [1,M \times N])$ pixels, (x_i,y_i) represents the range of the $2S \times 2S$ neighborhood in the image is the spatial location of the ith cluster center.

$$d_s = \sqrt{(x_j - x_i)^2 + (y_j - y_i)^2} \tag{2}$$

Formula (3) reflects the final distance including pixel distance and spatial distance, and m is a constant whose value range is generally$[1,40]$,this article also $m = 10$.

$$D\prime = \sqrt{\left(\frac{d_c}{m}\right)^2 + \left(\frac{d_s}{S}\right)^2} \tag{3}$$

Finally, the cluster center of each super pixel is recalculated, and the labels corresponding to the cluster center and pixel are constantly corrected. The calculation method of the clustering center as shown in formula (4), In the formula, (l_n,a_n,b_n) represents the new clustering center, (l_{jk},a_{jk},b_{jk}), γ represents the total number of k pixels labeled as.

$$\begin{cases} l_n = \frac{\sum l_{jk}}{\gamma} \\ a_n = \frac{\sum a_{jk}}{\gamma} \\ b_n = \frac{\sum b_{jk}}{\gamma} \end{cases} \tag{4}$$

Figure 2 shows the super pixel segmentation results ($S = 50, m = 10$) under different iteration times. It can be seen from Fig. 2 that with the increase of iteration times, the specific area of each super pixel becomes smooth, and its boundary coincides with the pixels in the picture.

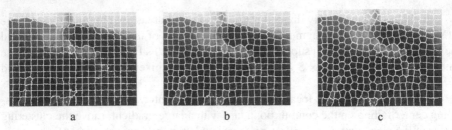

a   b   c

Fig. 2. Super pixel segmentation results of SLIC under different iterations.

2.3 Smoke Segmentation Method Based on SLIC and SegNet

From the above analysis, it can be found that SLIC can extract the clustering results between different pixels of the image, and SegNet can obtain the specific position of smoke pixels. In this paper, the smoke is further judged through the recognition results of SegNet and SLIC.

It is assumed that the smoke segmentation result of SegNet is a matrix. Since smoke segmentation is a binary classification problem in this paper, smoke pixels are denoted as $(A(i,j) = 1)$ and non-smoke pixels are denoted as $(A(i,j) = 0)$. The identification result of SLIC is denoted as a matrix, and the elements in the matrix correspond one-to-one to the labels of the SLIC identification result. Further processing of the judgment result of SLIC is carried out according to the form of formula (5).

$$S(k) = \frac{\sum_{i=1,j=1}^{M,N}(A(i,j) \wedge B(i,j) = 1)}{num_k} \tag{5}$$

The meaning represented by formula (5) is to solve the ratio between the number of pixels k belonging to smoke pixels in each label pixel k of SLIC and the value num_k of all pixels belonging to label pixels of SLIC. In other words, $S(k)$ represents the probability that each super pixel in the SLIC belongs to smoke. The smoke image can be segmented by extracting the super pixel that satisfies the specific value of $S(k)$. The larger the value of $S(k)$, the more pixels that contain smoke in the super pixel. Figure 2 shows the results of the $S(k) > 0$ and $S(k) == 1$ (Fig. 3).

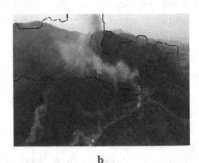

a   b

Fig. 3. The result of combining SLIC and SegNet (Figure a $S(k) > 0$ Figure b $S(k) == 1$)

Set the threshold variable as a, when (S(k) > α), the super pixel segmentation area is considered to be smoke, so that smoke can be judged by different threshold values a.

3 Experiment and Discussion

This dataset consists of a forest fireworks dataset, an aerial dataset, a video surveillance dataset, and an individual smoke dataset taken from the field. Contains forest fire smoke scenes, car fire smoke scenes, house fire smoke scenes, farm fire smoke scenes; The data set provides different shapes, different colors, and backgrounds of smoke, through the training of the data set, so that the computer can identify the smoke in different circumstances, so as to provide the most solid data guarantee for the early warning system. After a lot of training, it also makes the computer for different smoke characteristics to further understand, thereby reducing the error recognition rate.

3.1 Smoke Image Segmentation Results Based on SegNet

In this paper, the Deep Learning Toolbox™ and Image Label tools provided in MALTAB software were used to annotate the image and implement the SegNet algorithm. In the process of labeling photos containing smoke images, the area with dense smoke was emphasized, while the front edge of smoke was not labeled. Not all the smoke images labeled in this paper are taken by drones, and there are also a large number of samples from smoke photos taken on the ground. The SegNet neural network is trained through a large number of smoke photos, in order to obtain a better smoke segmentation effect of the model. In order to speed up the training speed and efficiency, this paper unified the size of all smoke images and smoke labels to the standard size of 300*400*3, and used the stochastic gradient descent method to train the neural network, with a momentum of 0.9 and an initial learning rate of 0.1.The specific results of classifying images containing smoke using SegNet are presented in Table 1.

Table 1. Smoke segmentation results based on SegNet

It can be seen from Table 1 that the smoke detected by SegNet is more accurate, but the details of the smoke part are not obvious. For example, there are incorrectly segmented pixels in serial number one picture, and some smoke that overlaps with the sky in serial number two picture is not segmented. On the whole, the SegNet pixel segmentation method can be used for semantic recognition of smoke in a single frame image, but some smoke pixels are missed or misdetected. The above phenomenon is mainly caused by the following two factors: first, some non-smoke pixels in the image are confused because they are similar to smoke pixels; Second: there is a labeling error in the process of labeling. It can also be found from Table 1 that there are many pixels in the smoke segmentation results using SegNet that are independent and not connected with other smoke pixels.

3.2 Super Pixel Segmentation Results Including Smoke Pixels

The results of super pixel segmentation are mainly directly related to the selection of, and the larger the value, the larger the area of super pixel. The relationship between different values and super pixel segmentation results and the specific time of super pixel recognition are given in Table 2.

Table 2. Super pixel segmentation results and segmentation time.

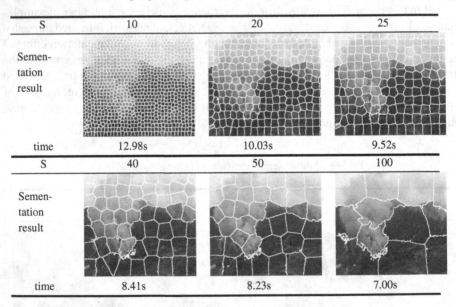

S	10	20	25
Sementation result			
time	12.98s	10.03s	9.52s
S	40	50	100
Sementation result			
time	8.41s	8.23s	7.00s

Can be seen from Table 2 super pixel segmentation results and have a close relationship between the size of the super pixel, super pixels, the smaller the details is divided, the more careful, the unit area can be found within the pixels can be clustering into more detailed a number of small, when the image is divided in the area of large span and super pixel contains more information, For example, the fourth super pixel block in the second row contains not only the elements of mountains but also some elements of smoke. Therefore, the smaller the super pixel, the more unitary the element information is.

The computer used in this paper is configured with intel 12 generation i7-12700FCPU, 32G memory, and GeForce RTX 2060 Super graphics card model. It can also be seen from Table 2 that the segmentation size of super pixels is inversely proportional to the segmentation time, and too small will lead to too long time of super pixel segmentation, which does not meet the actual engineering requirements. To sum up, in this paper, it is recommended to use the super pixel element which is relatively single and has a short operation time.

3.3 Smoke Segmentation Results Based on SLIC and SegNet Algorithms

The final smoke segmentation result depends on the threshold value. The larger the value, the stricter the smoke segmentation, and the smaller the value, the broader the smoke segmentation. Table 3 shows the smoke segmentation results under different conditions, in which the area surrounded by the black line is the area of smoke pixels.

Table 3. Smoke segmentation results under different thresholds. $S = 40$

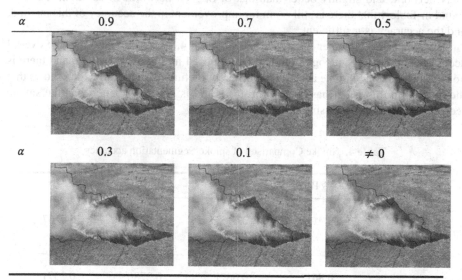

α	0.9	0.7	0.5
α	0.3	0.1	$\neq 0$

3.4 The Advantages of the Results of This Paper over other Methods

Observe whether smoke segmentation result is an important evaluation standard depends on its good accuracy. When the initial images of the smoke region segmentation and the original smoke regional coincidence degree is higher, it shows that the smoke has the better segmentation effect. The method is presented in Table 4 smoke pixel accuracy segmentation results compared with other methods.

The initial images and images with different methods are marked first, and then binary processing is carried out. The size of the images is uniformly set to 400*300, and then formula (6) is used for phase and calculation. Then formula (7) is used to calculate the ratio of all pixels with the result of 1 to get the final accuracy.

$$K1 = I\&J \tag{6}$$

$$Y = \frac{A}{400 * 300} \tag{7}$$

It can be seen from Table 4 that the accuracy of smoke segmentation results of different methods is quite different. In the segmentation results of the first image, the SegNet network has the lowest accuracy of smoke pixel segmentation, while the U-Net network has the highest accuracy. Although the accuracy of the proposed method is relatively lower than that of FCN network, the result difference is only 0.0188; It is not difficult to see from the comparison results of the accuracy of the second image that the accuracy of the proposed method in this paper is better than that of the other three methods, and the accuracy rate is significantly higher than that of U-net network and FCN network, and slightly better than that of SegNet network. In addition, it can be found in Table 4 that there is still room for improvement in the accuracy of this method, and the accuracy is insufficient.

By comparing the other three methods in Table 4, it can be found that the overall accuracy of the smoke recognition method proposed in this paper is high, and there is no significant change in the accuracy of different picture recognition, which shows that the method in this paper has better robustness and is generally better than the smoke recognition method only using deep learning.

Table 4. Smoke Comparison of smoke Segmentation accuracy

Result	This paper	SegNet	U-Net	FCN
	0.9136	0.6151	**0.9324**	0.9167
	0.8611	0.6671	0.2780	0.2571

It can be seen from Table 3 that the choice of threshold has a great influence on smoke segmentation. When the threshold is large, the results of smoke segmentation will be more rigorous, and the segmented region is the region with dense smoke. On the contrary, when the threshold is small, the result of smoke segmentation will be more relaxed, and the segmented region contains the region with sparse smoke. In practice, it is recommended to use a larger value when the smoke is relatively dense and the range is relatively wide, and the segmented area contains the main part of the smoke. It is also easy to see from Table 3 that some smoke details are still not segmented, such as the smoke pixels at the top left of the image, which is mainly because such light smoke is not labeled in the labeling process of smoke. Compared with the SegNet recognition results in Table 1, it is easy to find that the smoke recognition method proposed in this paper is better than the smoke recognition method using only deep learning. The details of the smoke edge are more obvious, and there are no independent pixels.

4 Conclusion

This paper proposes a single-frame image smoke segmentation method based on SLIC and SegNet. The method uses the SegNet algorithm to identify the pixels of smoke, determine the general outline and specific position of smoke, and then uses the SLIC algorithm to perform super pixel segmentation of the single-frame image containing smoke. Then, according to the recognition results of SegNet, the super pixels segmented by SLIC super pixels are assigned corresponding probabilities, so as to obtain the relationship between different super pixels and smoke. Finally, the probabilities in the super pixels are judged for different scenarios to determine the final smoke segmentation results. From the experimental results, the segmentation effect of smoke using the algorithm in this paper is better than that using only the deep learning algorithm, and the smoke edges and details segmented by the method in this paper are more clear and accurate.

References

1. Chen, T.H., Wu, P.H., Chiou, Y.C.: An early fire-detection method based on image processing. In: 2004 International Conference on Image Processing, 2004. ICIP 2004, Singapore, vol. 3, pp. 1707–1710 (2004). https://doi.org/10.1109/ICIP.2004.1421401
2. Phillips Iii, W., Shah, M., da Vitoria Lobo, N.: Flame recognition in video. In: Pattern Recogn. Lett. **23**(1–3), 319–327 (2002). ISSN 0167–8655
3. Wang, Z., et al.: Predicting transient building fire based on external smoke images and deep learning [J]. J. Build. Eng. **47**, 103823 (2022)
4. Abdusalomov, A.A., et al.: An improvement of the fire detection and classification method using YOLOv3 for surveillance systems [J]. Sensors, **21**(19), 6519 (2021)
5. Saponara, S., Elhanashi, A., Gagliardi, A.: Real-time video fire/smoke detection based on CNN in antifire surveillance systems [J]. J. Real-Time Image Proc. **18**, 889–900 (2021)
6. Larsen, A., et al.: A deep learning approach to identify smoke plumes in satellite imagery in near-real time for health risk communication [J]. J. Eposure Sci. Environ. Epidemiol. **31**(1), 170–176 (2021)
7. Khan, S., et al.: Deepsmoke: deep learning model for smoke detection and segmentation in outdoor environments[J]. Expert Syst. Appl. **182**, 115125 (2021)
8. Yar, H., et al.: Vision sensor-based real-time fire detection in resource-constrained IoT environments [J]. Comput. Intell. Neurosci. **2021** (2021)
9. Yuan, F., et al.: A gated recurrent network with dual classification assistance for smoke semantic segmentation [J]. IEEE Trans. Image Process. **30**, 4409–4422 (2021)

10. Dubey, V., Kumar, P., Chauhan, N.: Forest fire detection system using IoT and artificial neural network[C]. In: International Conference on Innovative Computing and Communications: Proceedings of ICICC 2018, vol. 1, pp. 323–337. Springer, Singapore (2019)

11. Avazov, K., et al.: Forest fire detection and notification method based on AI and IoT approaches[J]. Future Internet **15**(2), 61 (2023)

Research on Distributed Routing Technology for LEO Satellite Network

Zhongyu Yin[1], Cheng Zhang[2], Xi Wang[1], and Shuo Shi[1(✉)]

[1] School of Electronic and Information Engineering, Harbin Institute of Technology, Harbin 150001, Heilongjiang, China
crcss@hit.edu.cn
[2] Aircraft Design Research Institute in Shenyang of Aviation Industry, Shenyang 110000, Liaoning, China

Abstract. During the development of 6G networks, cellular networks that rely solely on ground infrastructure cannot meet network coverage in remote areas. On this basis, the concept of Low Earth Orbit (LEO) satellites presents a unique solution to the coverage problem of 6G networks in special areas. Depending on the forwarding function provided by satellites and inter-satellite links (ISL), the LEO satellite network can achieve global network coverage. However, due to the huge scale of satellite constellation, how to achieve low-cost and reliable routing in satellite network research has become a huge problem. Based on the characteristics of LEO satellite networks combined with service requirements, this paper analyzes the research status of LEO satellite network routing protocols.

Keywords: Satellite Communication · LEO · Routing protocol

1 Source and Purpose of the Research

With the progress of ground mobile communication networks and the popularity of 5G networks, 6G networks have become the focus of current research. In the process of 5G network construction, cellular networks that rely solely on ground infrastructure have also exposed many deficiencies in network coverage, such as the inability to provide network access in Marine areas and the low cost-efficiency ratio of building base stations in remote areas, deserts and other uninhabited areas [1]. These problems also need to be solved in 6G networks.

On the other hand, in recent years, with the progress of multi-satellite technology of carrier rockets at home and abroad and the rise of private commercial aerospace, thanks to the emergence of key technologies such as recoverable launch vehicles, Low Earth Orbit (LEO) satellites, In particular, the concept of LEO Mega-Constellations provides a unique solution to the 6G network coverage problem in specific areas [2]. The so-called low-orbit giant constellation satellite network refers to a class of satellite networks with an orbital altitude of 500-2000km and the number of intra-orbit satellites in the constellation reaching thousands or even tens of thousands of magnitude [3]. Due

J. Li et al. (Eds.): 6GN 2023, LNICST 554, pp. 263–274, 2024.
https://doi.org/10.1007/978-3-031-53404-1_24

to its orbital height, LEO satellite has the advantages of low launch cost, large network capacity, low delay and high coverage, which can be used for user network access [4] services in 6G networks, and make up for the shortcomings of ground networks with poor coverage in special areas.

However, while having many advantages, the huge number of LEO satellites and the topological characteristics of their networking and service provision also pose a big problem to the networking and routing technology. Specifically, the problems to be solved come from three aspects:

(1) Inter-star routing problem.

Traditional medium-high orbit satellites generally designate a fixed ground station for each satellite. However, LEO satellite network is different. Due to its low orbital altitude and small coverage area of a single satellite, multiple satellites are often required to carry out real-time multi-hop relay and forwarding through ISL to meet the needs of communication service coverage, which requires appropriate routing protocols to achieve.

(2) Dynamic topology problem.

The traditional medium and high orbit satellites generally have a relatively long operating period, and the position relationship is relatively constant when communicating with the ground or other satellites. But on the LEO satellite network, due to the low orbit, the satellite period is short, and the viewing duration of a single satellite is only on the order of minutes, which requires frequent switching of satellites, resulting in the highly dynamic satellite-earth topology of the LEO satellite network. On the other hand, since the relative positions of satellites with different orbits change rapidly when they are near the poles or at high latitudes, ISL may be interrupted or need to be re-selected. Therefore, the inter-satellite topology of LEO satellite network is also dynamic. Therefore, LEO networking also puts forward the need for routing protocols to adapt to dynamic topology.

(3) Distributed routing problem.

In general, the routing of traditional LEO satellites is highly dependent on the scheduling and control of MEO or GEO satellites with higher orbit altitude and larger coverage. This is a centralized routing method. The routing of packets between LEO satellites is calculated by higher-orbiting satellites that know all or most of the LEO satellites' intra-orbit status [5], which in effect assume the duties of the dispatch center.

From the point of view of damage resistance and cost, this centralized routing method relies heavily on the reliable operation of high-orbit satellites, and has poor damage resistance and high networking cost. Therefore, the distributed routing problem that does not rely on high-orbit satellites and ground stations, completely relies on each LEO satellite itself, and can adapt to the characteristics of giant constellations is also a problem that the routing protocol needs to focus on.

The above three aspects of low-orbit satellite networking reflect that the low-orbit satellite network meets the nature of ad-hoc network in technology, and the solution of these three problems is also the focus of Ad-hoc network routing protocol research. Therefore, this paper will analyze the requirements from the perspective of Ad-hoc networks, and study a self-organizing routing protocol based on Walker constellation for LEO satellites, especially the LEO giant constellation network, and solve the problems encountered in the current LEO satellite network with the idea of Ad-hoc networks.

2 Development Status of LEO Satellite Network

In recent years, due to the decrease of launch cost and the increase of communication demand, the number of communication satellites in orbit has exploded. Compared with the traditional GEO communication satellites, the communication satellite network of Nova has two distinct characteristics: on the one hand, the orbit altitude is low orbit, and the LEO orbit with an altitude of about 600 km is generally adopted; On the other hand, the scale of the constellation is huge, and the number of satellites in orbit generally reaches hundreds to thousands of orders. As a representative of this type of nova satellite network, the current LEO satellite network for the purpose of providing communications services is as follows:

2.1 Starlink

Starlink is a next generation of low-orbit satellite communication network proposed and constructed by Space X Company of the United States. Its original design is very representative in the low-orbit satellite network. Specifically, it has two points: first, it provides fixed network coverage for the areas where ground base stations and fixed networks are difficult to reach; The second is to provide direct end-to-end connectivity for devices accessing the Starlink network.

In addition, the Starlink generation of satellites does not have ISL for LEO inter-satellite communication forwarding, but the latest Starlink satellites have been equipped with laser ISL for satellite transport. The complete ISL networking communication function is expected to be launched in 2023, which will improve the reliability and capacity of ISL network connection [6].

2.2 OneWeb

OneWeb is a low-orbit satellite communications network built by a British company of the same name. OneWeb's design goals are similar to Starlink's, both of which can provide network access and cross-star communication services for users, but the biggest differences are reflected in the business approach and direct user base. OneWeb is primarily targeted at aviation, maritime, enterprise and government, and is not directly open to home users, but individuals can also purchase OneWeb's network access services indirectly through network service operators.

The satellites of the OneWeb network do not use ISL for interconnection from the beginning of design, do not have on-board processing and inter-satellite multi-hop forwarding functions, and can only bend the pipe for transparent forwarding of data between ground stations.

2.3 Kuiper

Kuiper is a low-orbit satellite communication network proposed by Amazon. Unlike Starlink and OneWeb, which are only used for Network coverage and end-to-end communication, the satellites in the Kuiper system are planned to use Software Defined Network (SDN) technology, such as satellites with full on-board regeneration, on-board switching, and on-board repackaging capabilities.

The planned Kuiper constellation consists of 3,236 LEO satellites at three orbital altitudes of 590, 610 and 630 km, with inclinations between 33 and 51.9°. The planned Kuiper satellite will use laser ISL to achieve inter-satellite communication, so as to achieve on-board transmission.

3 Topological Characteristics of LEO Satellite Networks

There are two decisive factors affecting the topological characteristics of LEO satellites [7, 8]. One is the satellite constellation, which determines the motion orbit of satellites and the position relationship between multiple satellites, laying the foundation for the on-off of ISL. The second is the connection and selection of ISL, which determines the specific structure of the final inter-star topology and ultimately affects the design of routing protocols. The following article will make a specific analysis of these two factors:

3.1 Satellite Constellation

According to the summary in Sect. 2.1, the existing LEO satellite system generally adopts the configuration of Walker constellation. Walker constellation is a uniformly distributed constellation, which is evenly reflected in two aspects: First, a complete Walker constellation has several circular orbits with the same inclination, and the Angle between each orbit and the intersection line of the equatorial plane is equal, that is, the orbital plane is evenly distributed; Second, there are several satellites in each orbit, and the Angle between each two adjacent satellites relative to the center of the earth is equal, that is, the satellites are evenly distributed in the orbit.

Due to the homogeneous nature of the Walker constellation, the position of each satellite in the constellation can be uniquely determined by a simple set of mathematical formulas. However, to study the mathematical description of Walker constellation, first clarify the description of satellite orbits.

Using the orbital roots, we can further derive the mathematical form of the Walker constellation, a Walker constellation can be determined by 2 orbital parameters and 3 configuration codes:

First, the two orbit parameters are orbit altitude and orbit inclination respectively.

Since Walker constellation uses a circular orbit, the eccentricity is constant, the perigee concept can be chosen as 0 (which means that the ascending intersection of each orbit is the 0° reference point of the near-point Angle), and the value of the semi-major axis of the orbit is equivalent to the orbit radius. The relationship between orbit radius and orbit height is as follows:

$$r = h + R \tag{1}$$

where $R = 6378\text{km}$ is the mean radius of the Earth. In conclusion, the semi-major axis, eccentricity, perigee argument and orbital inclination are determined by the two orbital parameters and circular orbit characteristics of Walker constellation.

Secondly, the configuration code of Walker constellation which is called *N/P/F* are: the number of satellites, the number of orbital planes, and the phase factor.

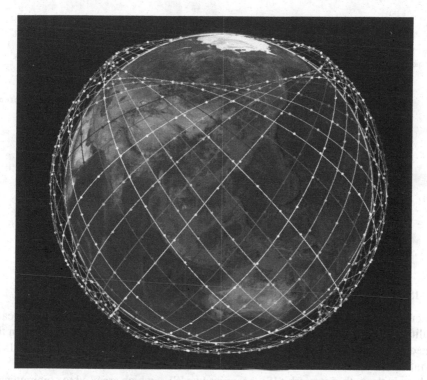

Fig. 1. Walker constellation simulation results.

The number of satellites is the total number of satellites that constitute a Walker constellation. Generally speaking, satellites are evenly distributed in each orbital plane, so the factor of is required. The phase factor is an integer used to determine the phase relationship between two adjacent orbital satellites. Specifically, with the help of the N/P/F configuration code, we can determine the ascension right ascension of a satellite and the near-point Angle at a certain time by the following formula:

$$\begin{cases} \Omega_i = \dfrac{360}{P}(P_i - 1) \\ \varphi_i = \dfrac{360P}{N}(N_i - 1) + \dfrac{360}{N}F(P_i - 1) \end{cases} \tag{2}$$

where P_i is the number of the orbit where the satellite is located, and the values is $P_i = 1, 2, \cdots, P$; N_i indicates the number of the satellite in the P_i orbit, and the value is $N_i = 1, 2, \cdots, N/P$.

Taking the single-layer Walker constellation with an orbital altitude of 630km, inclination of 51.9° and configuration code 1156/34/1 in the Kuiper constellation as an example, a complete Walker constellation is shown as Fig. 1.

According to the above properties of Walker constellation, the connection relationship and stability of ISL can be further analyzed.

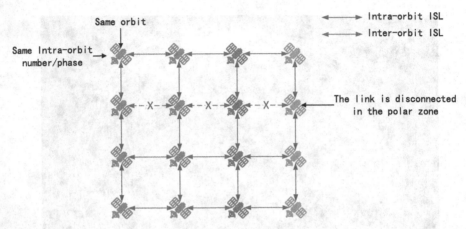

Fig. 2. Topology diagram between satellites.

3.2 ISL Connection and Selection

The current research on LEO satellite network routing generally assumes that each satellite has four ISLs for communication and service forwarding with satellites in the same constellation [9]. The four ISLs are divided into two categories: intra-track ISLs and inter-track ISLs.

Among them, the intra-orbit ISL occupies two ISLs, which are used to communicate with neighboring satellites in the same orbit. It can be expected that the ISL in the rail is relatively stable because the relative positions of both sides are fixed. However, inter-orbital ISL occupies the remaining two ISLs for communication with satellites in adjacent orbits. As the relative positions change in real time, the selection of inter-orbital ISL is a major problem in LEO networking. If appropriate, a relatively stable inter-satellite topology will be formed and the design requirements of routing protocols will be simplified, while if inappropriate, the topology will change rapidly. Increase the burden of routing protocols.

In recent years, there have been many researches on the influence of ISL selection on the topology. It is pointed out in the literature [10] that reasonable ISL selection can ensure that the ISL will not be disconnected in a complete operating cycle due to the abrupt change in the position relationship when the direction of elevation and elevation of the polar region changes. Although this choice of ISL would allow the interplanetary topology to be considered static and greatly simplify the design of the routing protocol.

topology probe part, it would give the interorbital lateral ISL in the inclined Walker constellation a helical topology, where the interorbital ISL does not return to the same satellite in its original orbit after one orbit around the Earth. This will have additional effects on the routing mechanism; However, the view mentioned in literature [9] is more general, which holds that the existence of inter-orbital ISL makes LEO satellite network present a standard grid topology, and inter-orbital ISL will be temporarily.

interrupted when satellites enter the polar region, and will be rebuilt when they leave the polar region. All kinds of ISL selection mechanisms also have higher stability and are more concise in the low latitude region, but as a cost, they need to put forward dynamic topological adaptability requirements for routing protocols, and need to predict and avoid ISL disconnection. Considering that the research purpose of this paper needs to adapt to link interruption caused by various reasons, the latter model is adopted in this paper, and it is considered that the inter-star topology can be abstracted as the grid shown in Fig. 2.

4 LEO Network Service and Routing Requirements

LEO satellite network can be divided into three networking modes according to the service model it carries.

4.1 Satellite Earth Network

The most classic communication satellite business model, also known as bent pipe forwarding. In this business model, the satellite assumes the transparent forwarding function, does not process or route the user's data on the satellite, and only forwards the user's terminal's data on the satellite to the ground station visible to the satellite at the same time.

From the link point of view, the function of satellite is essentially equivalent to bridging the user and the ground station that cannot establish a direct link through the method of relay. Therefore, from the perspective of providing users with network access services, the satellite does not need to undertake the routing function of the network layer, in fact, the user accesses the ground network provided by the ground station, so it is named the star ground network. According to the analysis in Sect. 2.1, OneWeb only provides such services in the existing LEO satellite network, and such services require the user and the ground station to be within the communication range of the satellite at the same time, so the orbital altitude of OneWeb constellation is higher than that of Starlink and Kuiper. Since there is no need for ISL and inter-satellite routing and forwarding, this business model is not within the scope of this paper.

4.2 Satellite-Based Network

In contrast to the space-based network service model in which the network functions are completely borne by the ground, the network functions are completely borne by the satellite. In this business model, a satellite-based satellite network independent of ground network facilities is formed by the ISL connection between satellites and the calculation by satellite travel. The user terminal can access the space-based network through the satellite-based link, and directly connect any device connected to the same space-based network without going through the ground network through the satellite-based route forwarding, so as to achieve the end-to-end connection of the satellite.

In terms of the demand for routing protocol, using the analogy of ground network, the networking communication between satellites in satellite-based network assumes the function of the core network, so the routing protocol needs to detect and maintain the network status of each node (satellite) and the topological connection relationship between nodes in real time, so the routing protocol also needs to complete the addressing and pathfinding functions of the specific terminal. The above requirements can be realized in the ground network through OSPF and other protocols at a low cost, but due to the dynamic topological characteristics of LEO satellite networks and the requirements of security and damage resistance, it needs to be optimized or redesigned.

4.3 Hybrid Network

The business model of hybrid network can be regarded as the synthesis of the business requirements of satellite Earth network and the technical means of satellite-based network. In this service model, on the one hand, satellites are no longer isolated from each other but connected through ISL, which can carry out inter-satellite networking and inter-satellite forwarding, breaking the coverage limit of bent-tube forwarding. On the other hand, the routing process is no longer completely limited to inter-satellite routing, and the destination host is no longer limited to the registered device of the space-based network, but the user's connection request can be forwarded to the ground gateway station through the inter-satellite relay, and the ground network can be connected through the gateway station to complete transmission and access.

The hybrid network service model combines the coverage area advantages of satellite network with the abundant bandwidth and rich resources of ground network, realizes the complement of the blind area of ground base station in 6G network, and completes the network coverage, which is the final design goal of existing LEO satellite network including Starlink and Kuiper. In terms of routing protocol requirements, in addition to topological detection and terminal addressing requirements in space-based networks, ground gateway stations may need multi-hop forwarding to reach and there are multiple alternative stations, so it is also necessary to optimize the routing function of ground gateway stations.

5 Routing Method for LEO Satellite Network

Compared with the ground mobile Ad-hoc network, the routing protocol for low-orbit satellite network, especially the giant constellation network, has two main features: First, the number of nodes participating in the network is large, and the cost of the simple flooding broadcast routing is too high; Second, although the topology changes dynamically, it still has regularity, which needs to be used.

In view of these two characteristics, there are generally two ideas to solve the dynamic topological routing problem of LEO satellite networks: virtual topology method and virtual node method.

5.1 Virtual Topology Method

Virtual topology method, also known as time virtualization, is a method to solve the problem of dynamic topological changes between satellites in satellite networks. Based on the regularity of the satellite network topology and the static characteristics of the topology in a short time, the time is discretized into continuous time slices, and the topology change only occurs when the adjacent time slices are switched, and the satellite network topology is fixed in each time slice. The static topology obtained in each time slice is named virtual topology because it does not reflect the topology of real satellite network in real time. The purpose of using virtual topology is two: First, route detection only needs to be carried out when the time slice changes, which can effectively reduce the route cost; Secondly, each satellite in each time slice can use static routing algorithm to calculate the route offline, which reduces the calculation overhead.

The theoretical basis of virtual topology method is the short-term stability of Walker constellation topology. Literature [11] studied this problem at the theoretical level and deduces the topological stability time for different number of satellites. However, it can also be found that virtual topology method is not suitable for giant constellation routing because of the high frequency of topology change and short time slice.

5.2 Virtual Node Method

Virtual node method, also known as space virtualization, is originally a solution to the problem of dynamic topological changes in satellite networks [28]. The idea is to divide the earth surface into different logical address areas according to the number of satellites and abstract each area into a virtual node, and then bind each satellite to the area where its ground projection is located. Finally, the route between the star/earth node and the star/earth node is transformed into a virtual node to virtual node routing problem. The purpose of using virtual node is two: First, virtual node has the actual geographical location, which can assist in route calculation; The second is the virtual node abstracts the satellite and the ground equipment connected to the satellite into the same node, which greatly reduces the routing cost.

The advantage of virtual node method is to transform the complex routing problem into the two-way routing problem of Manhatton grid type. However, the virtual node region division method originally proposed in literature [12] and the subsequent research optimization are completely based on the region division based on the standard of equal longitude and latitude under the condition of polar orbit Walker constellation. This division method is not suitable for the inclined orbit Walker constellation and needs to be further improved.

6 Routing Protocol for LEO Satellite Network

6.1 DT-DVTR

Discrete Time Dynamic Virtual Topology Routing (DT-DVTR) is a distributed link-state routing protocol based on virtual topology. The working process of DT-DVTR is divided into three steps: the first step is time slice. The protocol discretized the time according to the statistical analysis results of ISL state change and topological stability time, and determined the topology in each time slice according to the measurement and control results of the ground station. The second step is route calculation. After each satellite has mastered the topology of the satellite network in a local area, Dijkstra algorithm is used to calculate the route to the destination terminal according to the principle of the shortest physical path. The third step is route optimization. Considering that the route change between adjacent time slices may cause large delay jitter, the method of sliding time slice window or overall optimization of track period is adopted to modify the selected route to reduce the path change. The advantage of DT-DVTR is that the short-time strong correlation of LEO constellation topology reduces the routing cost caused by dynamic topology. The disadvantage is that there is no topology detection mechanism in the protocol, and the topology detection is completed by the ground station, so it cannot get rid of the control of the ground to achieve the real distributed routing.

6.2 OPSPF

Orbit Prediction Shortest Path First Routing (OPSPF) is another distributed link state routing protocol based on virtual topology. The working process of the protocol is divided into three steps: the first step is to obtain the satellite position. Each satellite independently obtains the position information of the LEO constellation satellite from the GEO satellite; The second step is topological prediction, which predicts the topology in the next time slice according to the current position state of the constellation. The third step is route calculation. The shortest path algorithm in OSPF is used to calculate the routes in the current time slice. From the perspective of virtual topology, OPSPF protocol can also be regarded as an improvement of DT-DVTR, which has two characteristics relative to the latter: First, it is not necessary to know the topology of the entire LEO satellite network in the current time slice from the ground station or high-orbit satellite, but only needs to know its own position and calculate the real-time state of the constellation combined with orbit parameters, which reduces the transmission and storage overhead of the routing table but cannot cope with link interruption and topology changes outside the statistical law.

6.3 DRA

The Datagram Routing Algorithm (DRA) is a distributed distance vector routing algorithm based on the virtual node method. DRA algorithm and its improved algorithm are completely based on the virtual node method, dividing the earth surface and the satellite and network equipment on it into virtual nodes according to the checkerboard grid and then abstracting them into virtual nodes, and calculating the next hop route to the target virtual node for each data packet independently and hop by hop on each satellite. On this basis, the DRA algorithm is divided into three steps: the first step is to estimate the direction of the shortest hop route according to the difference between the geographical position of the DRA algorithm and the destination virtual node, and estimate the direction of the shortest hop route based on the Manhattan grid characteristics of the inter-star topology under the assumption that the length of all inter-star links is equal; The second step is direction enhancement. The physical length of each link is estimated according to the real-time attitude of the current constellation, the direction of the next hop is optimized with the minimum propagation delay as the constraint condition, and the previous result of the shortest hop number is retained as the alternative route. The third step is congestion processing. The alternative route direction is selected according to the actual link load and interruption. DRA algorithm is the basis of the following routing protocol based on Walker constellation virtual node method.

6.4 LAOR

Location Assisted On-Demand Routing (LAOR) is a distributed on-demand distance vector routing protocol based on the virtual node method. LAOR protocol can be regarded as an improved version of AODV routing protocol in ground Ad-hoc networks for Walker satellite networks, which continues the idea of on-demand routing in AODV and utilizes the topological characteristics of Walker constellation and the characteristics of destination virtual node orientation known by virtual node method. Specifically, the LAOR routing process is divided into four steps: The first step is to estimate the request area. When the routing demand is generated, the source node estimates the existence area of the possible path according to the relative position of the target node and itself and generates the request area information; The second step is route detection. The source node broadcasts a routing request (RREQ) message, which contains the request area information to limit the broadcast range and reduce the routing cost. The third step is route reply. When the destination node receives an RREQ packet, the original route reply (RREP) packet establishes a route from the source node to the destination node. The advantage of LAOR protocol is that compared with DRA protocol, the route detection mechanism is added, so the routing result is more reliable and can adapt to irregular topology. However, the disadvantage of LAOR protocol is that it fails to take advantage of the topology change law of Walker constellation, so that the complete route detection must be carried out again when the topology change within the law occurs, and the routing cost is still not ideal.

7 Conclusion

This paper takes the routing problem of LEO satellite network as the goal, analyzes the orbital characteristics and topological characteristics of LEO satellite network, and on this basis, analyzes the advantages and disadvantages of existing routing protocols combined with the service requirements and service models of LEO satellite network. It is concluded that the existing routing protocols are not fully applicable to the inclined orbit Walker constellation.

Acknowledgement. This work is supported by the National Natural Science Foundation of China under Grant 62171158.

References

1. Attar, H., Abass, E.S., Al-Nairat, M., Alrosan, A.: Convolutionally coded wideband code division multiple access in 6G LEO satellite communication system. In: 1st International Engineering Conference on Electrical, Energy, and Artificial Intelligence, EICEEAI 2022. Institute of Electrical and Electronics Engineers Inc. (2022)
2. Xie, H., Zhan, Y., Zeng, G., Pan, X.: LEO mega-constellations for 6G global coverage: challenges and opportunities. IEEE Access **9**, 164223–164244 (2021)
3. Pardini, C., Anselmo, L.: Effects of the deployment and disposal of mega-constellations on human spaceflight operations in low LEO. J. Space Saf. Eng. **9**, 274–279 (2022)
4. Jia, Z., Sheng, M., Li, J., Niyato, D., Han, Z.: LEO-satellite-assisted UAV: joint trajectory and data collection for internet of remote things in 6G aerial access networks. IEEE Internet Things J. **8**, 9814–9826 (2021)
5. Hsu, Y.H., Lee, J.I., Xu, F.M.: A deep reinforcement learning based routing scheme for LEO satellite networks in 6G. In: IEEE Wireless Communications and Networking Conference, WCNC. Institute of Electrical and Electronics Engineers Inc. (2023)
6. Chaudhry, A.U., Yanikomeroglu, H.: Laser intersatellite links in a starlink constellation: a classification and analysis. IEEE Veh. Technol. Mag. **16**, 48–56 (2021)
7. Wang, C.-J.: Structural properties of a low Earth orbit satellite constellation - the Walker delta network. In: Proceedings of MILCOM '93 - IEEE Military Communications Conference, vol. 3, pp. 968–972 (1993)
8. Royster, T., Sun, J., Narula-Tam, A., Shake, T.: Network performance of pLEO topologies in a high-inclination walker delta satellite constellation. In: IEEE Aerospace Conference Proceedings. IEEE Computer Society (2023)
9. Han, Z., Xu, C., Zhao, G., Wang, S., Cheng, K., Yu, S.: Time-varying topology model for dynamic routing in LEO satellite constellation networks. IEEE Trans. Veh. Technol. (2022)
10. Werner, M., Frings, J., Wauquiez, F., Maral, G.: Topological design, routing and capacity dimensioning for ISL networks in broadband LEO satellite systems. Int. J. Satell. Commun. **19**, 499–527 (2001)
11. Wang, J., Li, L., Zhou, M.: Topological dynamics characterization for LEO satellite networks. Comput. Netw. **51**, 43–53 (2007)
12. Mauger, R., Rosenberg, C.: QoS guarantees for multimedia services on a TDMA-based satellite network. IEEE Commun. Mag. **35**, 56–65 (1997)

Analysis of Weld Pool Characteristics in Narrow Gap GTAW Welding Based on Passive Vision

Zehao Li[✉], Su Zhao, Wang Zhang, Lizhe Fan, and Jianhang Zhou

Shanghai Dianji University, Shanghai, China
1324827990@qq.com

Abstract. In the process of narrow gap pipeline welding, the welding quality can be controlled and judged by the characteristic information of the weld pool. The traditional active vision laser is limited by the geometric characteristics of the narrow gap groove, which cannot be projected into the workpiece groove, and the weld seam is easily blocked by the welding wire, which cannot obtain the weld seam characteristics, so the welding quality cannot be judged online. In this paper, the molten pool image is obtained directly by passive vision, and the complete contour of narrow gap welding molten pool is restored by means of median filtering noise reduction, Canny edge detection, area filtering screening and circle fitting. Two key characteristic parameters, curvature radius and advancing angle of molten pool contour, which can be used to characterize the welding quality, are extracted. The results of pipeline welding experiments show that the greater the welding current, the greater the radius of curvature and the forward angle of the molten pool profile. The variation range of characteristic parameters under different penetration states is analyzed, which provides a basis for subsequent parameter adjustment and matching.

Keywords: Narrow gap welding · Passive vision · Image processing · Molten pool characteristics

1 Introduction

In narrow gap welding, weld quality control mainly depends on the welder's visual subjective judgment of welding information. The construction environment and welder's state can easily lead to welding defects. The automatic welding equipment equipped with visual sensing can observe the shape of the front molten pool during welding, realize the real-time adjustment of welding parameters and improve the welding quality.

The visual image of the welding process contains many useful information such as weld, molten pool and surrounding. If the edge information of the molten pool con-tour can be accurately and quickly detected and the relationship with the welding process parameters can be established, it is of great significance for penetration control and welding quality. The active vision sensor has high precision and high reliability, and is widely used in various welding sites, which greatly improves the automation and accuracy of

J. Li et al. (Eds.): 6GN 2023, LNICST 554, pp. 275–281, 2024.
https://doi.org/10.1007/978-3-031-53404-1_25

welding [1]. Although this method has good anti-noise ability, it has the disadvantages of high price, inability to track complex joints and pre-check errors. Passive vision directly collects images of the welding area to obtain more intuitive information. Because there is no position error and time lag, this method can monitor the weld pool in real time. Lee et al. proposed an automatic seam tracking method for pulsed laser edge welding without auxiliary light source. The visual sensor directly observes the information of the weld and the molten pool [2]. However, only when the welding light is weak can a clearer image be obtained and the extracted features are not obvious. Similarly, Chen et al. also tried to use the passive vision system to solve the seam tracking problem of metal sheet butt arc welding [3], and studied an improved edge detection algorithm, but only when the welding current is small and the arc light is weak can obtain a clear image. However, in the study of narrow gap welding, only a small number of people pay attention to the extraction of effective information such as weld pool.

For narrow gap pipeline welding, it is impossible to use laser projection to extract information on the surface of the workpiece like an active vision system, so passive vision is used to directly obtain the information of the welding area. And because the welding wire is on the opposite side of the welding torch, the weld seam is easily blocked by the welding wire and cannot be further studied, so the study of the molten pool is particularly important. Therefore, this paper redefines the parameters of the molten pool, and designs the image processing algorithm to extract the characteristics of the molten pool, which is used as an index to study the relationship between them under different welding currents, and optimizes the matching of each parameter, so as to obtain a good weld forming to improve the welding quality (Fig. 1).

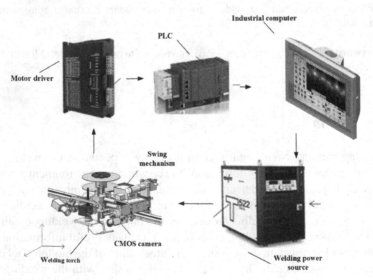

Fig. 1. Molten pool monitoring visual sensing system

The test platform of welding seam and weld pool monitoring video welding equipment mainly includes camera, welding power source, welding motion mechanism, controller, console, track and manual control box. The schematic diagram of the test platform is as follows:

2 Image Processing of Molten Pool

2.1 Extracting Feature Parameters from Molten Pool Images

In order to collect images of the molten pool during the welding process, the geometric shape parameter information of the molten pool is obtained after processing the molten pool image, and then the approximate relationship between the molten pool and the welding parameters is established. After graying out and median filtering, the molten pool image is then segmented by threshold to obtain the highlight and determine the center position of the molten pool. The size of the molten pool is then determined by threshold segmentation, and the curvature of subsequent ripples is determined by the center position, which corresponds to the actual welding situation.

2.2 Extracting Feature Parameters from Molten Pool Images

With the development of computer vision science and technology, the use of industrial camera head to realize the real-time acquisition of molten pool graphics, as well as the analysis of the relationship between molten pool morphological characteristics and weld parameters, has become a key scientific research direction for monitoring weld quality. In this paper, the weld pool image collected based on the passive vision system is shown in Fig. 2. After grayscale, it is shown in Fig. 3.

Fig. 2. Image graying of molten pool   **Fig. 3.** Median filtering

2.3 Threshold Segmentation Extracts the Center Point of the Weld Pool and the Weld Pool Image

Gray threshold segmentation is to divide the image gray into different levels, and then determine the gray, gate threshold method. Gray threshold segmentation is actually binarization processing, that is, selecting a threshold to convert the image into a black and white binary image for image segmentation and edge extraction. Image thresholding is a step function, which belongs to the nonlinear operation of image gray level. The curve of the transformation function is shown in the figure. Its function is to specify a threshold by the user. If the gray value of a pixel in the image is greater than the threshold, the gray value of the pixel is set to 255, otherwise the gray value is set to 0. The center point of the molten pool is obviously the brightest place in the whole image, so the center point of the molten pool can be segmented by threshold segmentation, and the threshold segmentation is shown in Fig. 4.

After obtaining the center point of the molten pool, the specific location of the molten pool can be located, reducing the perceptual area, and performing threshold segmentation again to reduce the value of the threshold segmentation. The resulting molten pool image is shown in Fig. 5.

Fig. 4. Initial threshold segmentation   **Fig. 5.** Secondary threshold segmentation

2.4 Extraction of Welding Patterns

The welding patterns vary in different welding situations. For example, as shown in the above figure, the patterns are relatively flat during normal welding. However, when the current is too high and the wire feeding speed is too fast (as shown in the following figure), the patterns after the molten pool are more wavy. Therefore, the curvature of the ripples after the molten pool can be used to determine whether it is a normal welding situation.

When the central brightness point is known, the sensitive area can be reduced and the ripple area on the left side of the molten pool can be selected. First, the edge is extracted by the Laplace operator, and then the ripple on the left side is screened by threshold segmentation and line screening, as shown in Fig. 6 and Fig. 7.

It can be seen that in the case of very high current, the curvature of the ripple behind the molten pool is much greater than in the case of very low current.

Fig. 6. Low welding current   **Fig. 7.** High welding current

2.5 Melt Pool Images Under Different Welding Currents

According to the definition of weld pool parameters, the characteristic parameters of the weld pool vary under different welding conditions. The higher the welding current, the faster the speed. The larger the weld pool, the larger the curvature radius of the weld pool contour, and the greater the difference between the start and end angles of the fitting circle. The inner and outer deviation will cause the average value of the start and end angles of the fitting circle to lean towards the opposite side. Therefore, the data obtained from the molten pool can be used to determine in reverse whether the system is in a situation where the current is too high, the wire feeding speed is too fast, and there is a deviation (Fig. 8).

Using the curvature radius (K) and the starting angle (θ_1) and ending angle (θ_2) of the fitting circle as characteristic parameters to characterize the molten pool, calculate the characteristic parameters of a total of 60 images to be processed in 7 experimental groups as shown in the following figure, and then record the average values in Table 1.

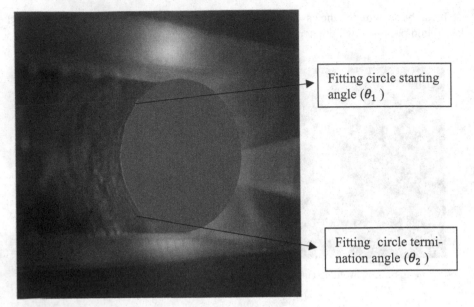

Fig. 8. Molten pool contour fitting.

Table 1. Characteristic parameter

Comparison of welding tests	Curvature radius (K)	angle (θ_1)	angle (θ_2)
Normal welding	0.00120926	169.765	182.246
High current welding	0.00368698	134.159	225.266
Low current welding	0.000713322	186.58	196.957

3 Conclusion

A passive vision based sensing system was built to directly obtain sequence images of the molten pool during narrow gap welding. In response to the difficulty in extracting molten pool features due to factors such as arc light and electromagnetic interference, median filtering denoising, Canny edge detection, and area filtering were used to extract the contour edges of the molten pool during the welding process.

In response to the problem of incomplete display of the molten pool in narrow gap welding, a circular fitting segment was used to calculate and restore the complete contour, and the curvature radius and forward angle of the contour were redefined as parameters to characterize the characteristics of the molten pool.

Through experimental comparison and analysis, the defined characteristic parameters of the molten pool can present the characteristics of the molten pool morphology. The influence of different welding current changes on the molten pool morphology was studied through pipeline welding experiments. The mapping relationship between current parameters, melt pool characteristics, and welding quality was obtained, which can provide a basis for parameter adjustment of penetration control in narrow gap pipeline welding.

References

1. Katayama, S.: Handbook of Laser Welding Technologies. Woodhead, Cambridge (2013)
2. Lee, S.K., Na, S.J.: A study on automatic seam tracking in pulsed laser edge welding by using a vision sensor without an auxiliary light source. 21(4), 302–315 (2002)
3. Chen, S., et al.: Computer vision technology for seam tracking in robotic GTAW and GMAW. Robot. Comput. Integr. Manuf. Int. J. Manuf. Prod. Process Dev. 32(C), 25–36 (2015)

Author Index

A

Aoki, Toya   I-3, I-10

B

Banban, Lu   II-109

C

Cai, Cheng   I-30, I-44
Cao, Junye   II-217
Chang, Liu   I-299
Chang, Zixuan   I-390
Chen, Zheyi   I-165
Chen, Zhimin   II-200
Chen, Zijia   I-117, I-129
chengkun, Wang   II-252
Chi, Dongxiang   II-217
Chunlei, Ji   II-208
Cui, Zaixing   I-363

D

Dai, Yingyu   II-188
Dou, meng shen   II-62
Dou, Nan   I-91, I-155
Du, Weian   I-353

F

Fan, Lizhe   II-275
Fan, Yi   I-390, I-409

G

Gao, Yang   I-30
Guo, Chunlei   I-91, I-155
Guo, Xu   II-96, II-160
Guo, Yanbing   I-257, I-267, I-277, I-347

H

Han, Jingxuan   II-217
He, Jinrong   II-3, II-18

Hong, Liu   II-224
Hou, Yichen   I-165
Huo, Xingdong   I-69

J

Ji, Chunlei   I-53
jiale, Yang   II-252
Jiang, Chao   II-31
Jiang, Hang   I-69
Jiang, Weiting   I-390
Jiangbo, Wu   I-214, II-224
jinqiu, Zhang   II-252

K

kaiyue, Feng   II-252

L

Leng, Peng   I-353
Li, Donghua   II-43
Li, HaoJie   II-160
Li, Haojie   II-96
Li, Jiacheng   II-149
Li, Jiandun   I-299, I-315
Li, Jiangtao   I-285
Li, Jingchao   I-105, II-188
Li, Kuixian   I-69
Li, Minzheng   II-43, II-52, II-200
LI, Tengfei   I-277
Li, Xiangyan   I-363, I-375
Li, Xinlu   II-175
Li, Yongguo   I-375
Li, Zehao   II-275
Lian, Zhigang   I-91, II-126
Liang, Xin   I-30
Liao, Rongyang   I-53
Liao, Wei   I-267, I-277, I-347
Lim, Hun-ok   I-3, I-10
Lim, Hun-Ok   I-17, I-23
Lin, Yun   I-79

J. Li et al. (Eds.): 6GN 2023, LNICST 554, pp. 283–285, 2024.
https://doi.org/10.1007/978-3-031-53404-1

Printed in the United States
by Baker & Taylor Publisher Services